POW

DE SNUIF M

Mural tablet:
Dutch tower mill.

ALTON MILL,
MUSEUM OF EAST ANGLIAN LIFE, STOWMARKET, SUFFOLK.

Power before STEAM

JOHN VINCE

JOHN MURRAY

POLDER MILL·
N.O.M. ARNHEM·

© JOHN VINCE   1985
First published 1985
by JOHN MURRAY (Publishers) Ltd.
50 Albemarle Street, London W1X 4BD

Printed and bound in Great Britain
by Fletcher & Son Ltd., Norwich.

British Library CIP Data

  Vince, John
    Power before steam
    1. Machinery - History
    I. Title
      621.8'09 TJ15
      ISBN 0-7195-4175-1

CONTENTS

C19 HANDCART USED TO
CARRY FUEL TO THE FIRES
AT ST. CROSS HOSPITAL,
WINCHESTER, HANTS.

INTRODUCTION

When MAN first encountered the elements he had to rely upon the power of his arm. After a long interval as a gatherer of nature's bounty he began to fashion tools for hunting, building and the first forms of agriculture. That enterprise set him apart from all other creatures and made him a technologist.

English hillsides still bear the marks made by our first farmers. Today's mapmakers refer to such prehistoric handiwork as 'Celtic Fields'. Within their narrow boundaries man tilled the soil with elementary scratch ploughs ~ see OLD FARMS ~ pulled by oxen. After man had subdued the ox it took him a long time to acquire the services of equus, and during the early mediaeval period it was the humble ox that provided draught for the plough. Rudyard Kipling in his 'Just So Stories' made the horse man's 'first servant'. From Tudor times onwards the horse played an increasingly important role in supplying man with a source of power. For centuries animal power had an important place in technical history, and even in the C19, after steam power was introduced, the plodding horse still had a role in workaday life.

Centuries probably passed after the first ox was yoked before some anonymous genius set the wind to work FOR instead of AGAINST man's designs. From Egypt we derive some of the earliest evidence (circa 3000 B.C.) which shows that wind was used to move its elegant ships along the Nile and out to sea. Once the principle had been applied, however, the wind was then enrolled to grind the harvest. Before this step another pioneer - the Roman Vitruvius usually gets the credit (C1. B.C.) - had devised a water-driven wheel to grind corn into flour. It is interesting to reflect that although the wind seems to have been the first element set to work for man, it was overtaken by water power which reigned as the most significant prime mover for some 1500 years.

Gravity as every schoolboy knows was 'discovered' by the famous Isaac Newton in his Lincolnshire garden at Woolsthorpe ~ 1666. Long before this, weight-driven machines had been devised to help man count the day's hours.

The evolution of cast iron in the C18 added significantly to the millwright's repertoire and it allowed engineers to develop the next prime mover ~ STEAM.

There is little documentary evidence of our early technical history. Much trial and error probably preceded the craftsman's perfection of a device which worked in the required way and needed a minimum of servicing. One reason is that society in times past considered the work of the artisan to be of little consequence. But without the craftsman's innate intelligence and his corpus of wisdom received by word of mouth society would have been bereft of its wealth.

In the pages below the manner in which man has used his own muscle power, and power derived from animals, water, wind and gravity are illustrated. During the last twenty years a new interest in our technical heritage has developed. Each year thousands of visitors re-live a fragment of our ancestors' lives by going to those places which were once a part of the industrial landscape, and a real part of our forbears' daily toil. Manufactories were almost always dirty, noisy, too hot or too cold, often dangerous and something to be endured for long exhausting hours. William Blake's "dark Satanic mills" do represent one aspect of industrial society, but this is not the only view. Some industrialists like Samuel Greg at Quarry Bank, Styal, Cheshire, created a different kind of industrial community. He provided his workers with a roof over their heads, a creche, a school and places of worship. Many of the examples discussed are accessible to the public ~ see p. 154.

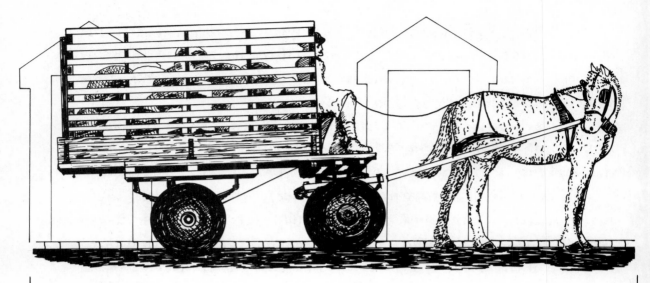

HORSE POWER is not a thing of the past. High motor fuel costs have once more made the horse a viable alternative for town deliveries. In Bristol the FRIENDS of the EARTH use this horse-powered unit to collect waste paper for recycling.

There were doubtless some ways in which elementary forms of power were used that are not mentioned here. Readers who are aware of examples that the author has left out are invited to tell him. It is important to record details of the different craft processes before they are lost. Even though so much has been done to make a record there are still many nooks and crannies to be explored.

Many individuals now devote their lives to caring for and operating corn mills. Their reward is not measured in financial terms. Rather it is manifest in the dignity, devotion and serenity that is derived from working with a known and fundamental source of power.

MAN

From the beginning man had to rely upon his muscle power in his struggle to master his environment. It seems likely that one of his first rudimentary tools was a stick. At some stage he discovered that it could be used to move heavy objects more easily, and so he began to unroll the mysteries of mechanics. Although the theory of levers took a lot longer to be enshrined in a scientific form prehistoric-pragmatic man put it to practical use. From his working knowledge of levers came one of the earliest machines – the windlass. This device was used to hoist sails on very early ships, and it was employed for many other tasks as well. In this chapter we shall look at some of the ways in which man generated movement and power by using his own limbs.

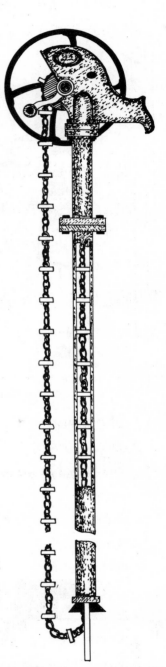

HAND OPERATED CHAIN-BUCKET PUMP C19.
Called a YEDDLE PUMP in the north of England.

THREE-POSITION CAPSTAN for
hauling boats up a slipway.

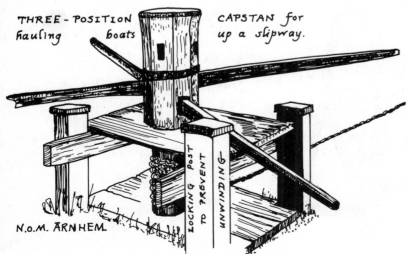

LOCKING POST TO PREVENT UNWINDING

N.O.M. ARNHEM

THE POLE LATHE

We do not know who first had the idea of using the natural springyness of a tree to provide power for a lathe. The concept was understood and used in the C14.

Craftsmen, like the Bodgers of the Chiltern Hills, still used this elementary machine as late as the 1950s to turn chairlegs for the High or Chepping Wycombe furniture trade.

A pole lathe uses the strength of the tree to rotate a work-piece set between fixed centres. The power is controlled by a treadle linked to the pole by a rope. This rope is twisted round the workpiece. As the operator pushes the treadle downwards so the

workpiece rotates towards the chisel which shapes it. When the pressure on the treadle is released the spring of the tree pulls the treadle upwards ready for the next downward push of the operator's foot.

HAND & FOOT POWER

The treadle lathe was developed by the addition of a flywheel which helped to provide a more constant speed. The example shown here is of special interest as it belonged to John Smeaton, F.R.S. (1724-1794) the engineer who made a special study of wind & watermills. The treadle operates a crank axle with a flywheel. A cord around the flywheel turns the pulley on the lathe and so rotates the workpiece.

Another form of lathe which was to be found in hundreds of wheelwrights' workshops is shown opposite. The flywheel was mounted in its own heavy frame and could be worked by one or two boys. A rope around its circumference turned a pulley on the substantial lathe & this motion revolved the heavy elm blocks which were shaped to make the hubs - naves - of the wagon wheels. The task of turning the flywheel must have been tedious and hot work in summertime. Some examples of flywheels placed in the roof of a building are known - see Salaman, R. op. cit.

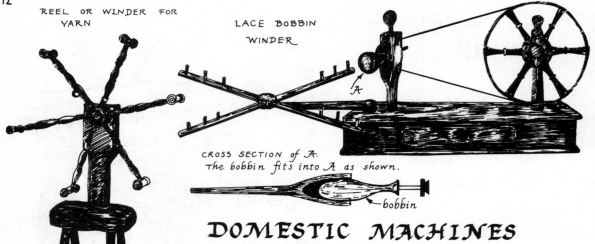

REEL OR WINDER FOR
YARN

LACE BOBBIN
WINDER

CROSS SECTION of A.
The bobbin fits into A as shown.

bobbin

DOMESTIC MACHINES

Technology began to affect the life of the housewife at a very early date. The first domestic machine was the spinning wheel which has a European ancestry going back to the C13. These wheels were hand-driven. To aid their momentum they were quite large ~ about 30" dia. Later on, C15, a treadle was added to this ancient tool, and the spinster at last had both hands free to apply to the spinning.

Lacemakers too had a need for a machine to wind their bobbins. These small table-top bobbin winders are now collectors' items.

The reel was used to wind spun yarn into hanks ready for the hand loom.

TREADLE SPINNING WHEEL

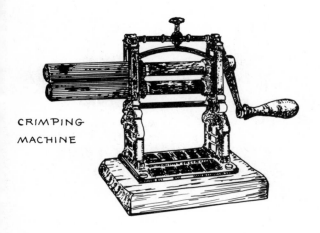

CRIMPING
MACHINE

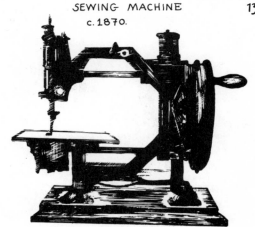

SEWING MACHINE
c. 1870.

In the Victorian and Edwardian period collars, caps and cuffs were given a crinkled finish. This crimping could more readily be achieved with the help of a machine which looks like a small mangle. Its brass rollers had a fluted surface.

The machine which probably made most impact on the C19 home was the sewing machine. Isaac Singer patented his first machine in 1851. The first examples were hand-operated but later versions had treadles. Early machines are collected by needlework enthusiasts. It is sufficient commentary on their workmanship & design to say that most will still work.

A less precise, but very useful, machine was the sausage stuffer, which used its gears to compress the meat & discharge it via the spout at the base.

The wooden American apple peeler has just two metal parts - the fork which holds the apple and the cutter A.

APPLE
PEELER

SAUSAGE
MACHINE

WASHING

During the C19 machines designed to reduce the hard labour that accompanied every washday began to appear. They agitated the washing. Example A is a rocking machine with features borrowed from the rocking chair. The second wooden machine B is a barrel with a horizontal arm attached to the central spindle. In use the arm was moved from side to side. Modern machines still copy this action. Machine C represents an advance on the design of the earlier prototypes.

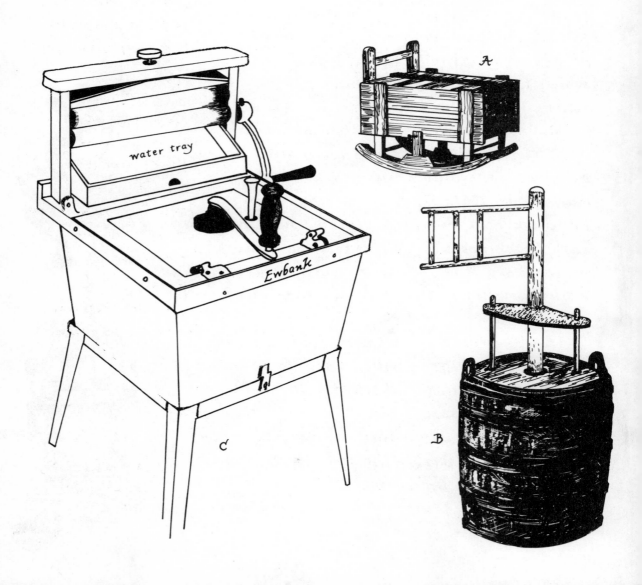

water tray

Ewbank

A

B

C

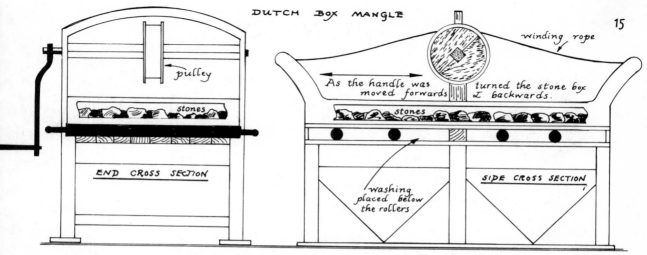

DUTCH BOX MANGLE

pulley

stones

END CROSS SECTION

winding rope

As the handle was moved forwards — turned the stone box & backwards.

stones

washing placed below the rollers

SIDE CROSS SECTION

It has a wringer attached and water squeezed from it ran down the water tray, back into the tub. In contrast to the older machines the Ewbank ~ c. 1930 ~ is made of steel sheet on angular metal legs.

Before wringers were invented the Dutch launderer twisted wet linen between a fixed post and a large hand wheel. Indoors the housewife probably began with a simple wooden roller and a mangle bat. Garments were wound round the roller and it was then rolled to and fro on a hard surface with the bat. The largest machine contrived to squeeze water from laundry was the BOX MANGLE. It had several wooden rollers which ran over the garments spread on the table. The weight of the stones in the box above the rollers provided the power which removed the water. As the handle of the mangle was turned the box moved backwards and forwards. This action moved the rollers across the washing.

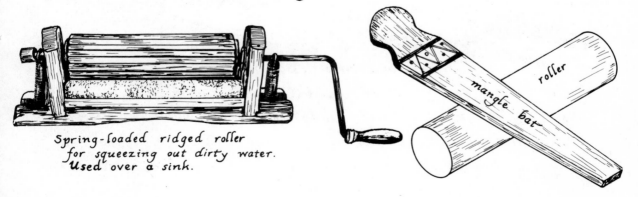

Spring-loaded ridged roller for squeezing out dirty water. Used over a sink.

mangle bat

roller

POTTERY

SLIP CART

Every aspect of industrial life once had a strong element of manual work. It is sometimes surprising to find just how long manual methods survived. At Etruria, that shrine of Wedgwood's consummate art, slip was still being moved about the works by hand even in the 1920s. Tubs of slip, the mixture of coloured clay used for decoration, were moved in the cart shown above. The tubs had a trunnion on each side which fitted into the forked end of the cart. The cart's long shaft was a useful lever to lift the heavy tubs and carry them away.

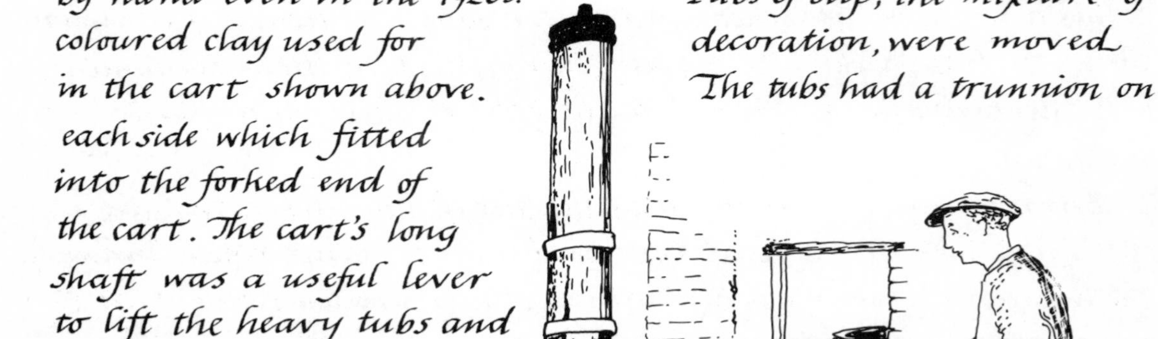

After James
Hodgkiss
1920

There were some tasks that could be performed better by hand. Mixing slip needed more than mechanical effort. The operator's eye was essential to tell exactly when the right mix had been achieved. This vat has its stone walls secured by iron straps. The inlet and outlet channels are on opposite sides.

PAINT GRINDING

Until the C19 most paint was made by reducing cakes of colour to a powder with a pestle & mortar. Linseed oil & lead was then added.

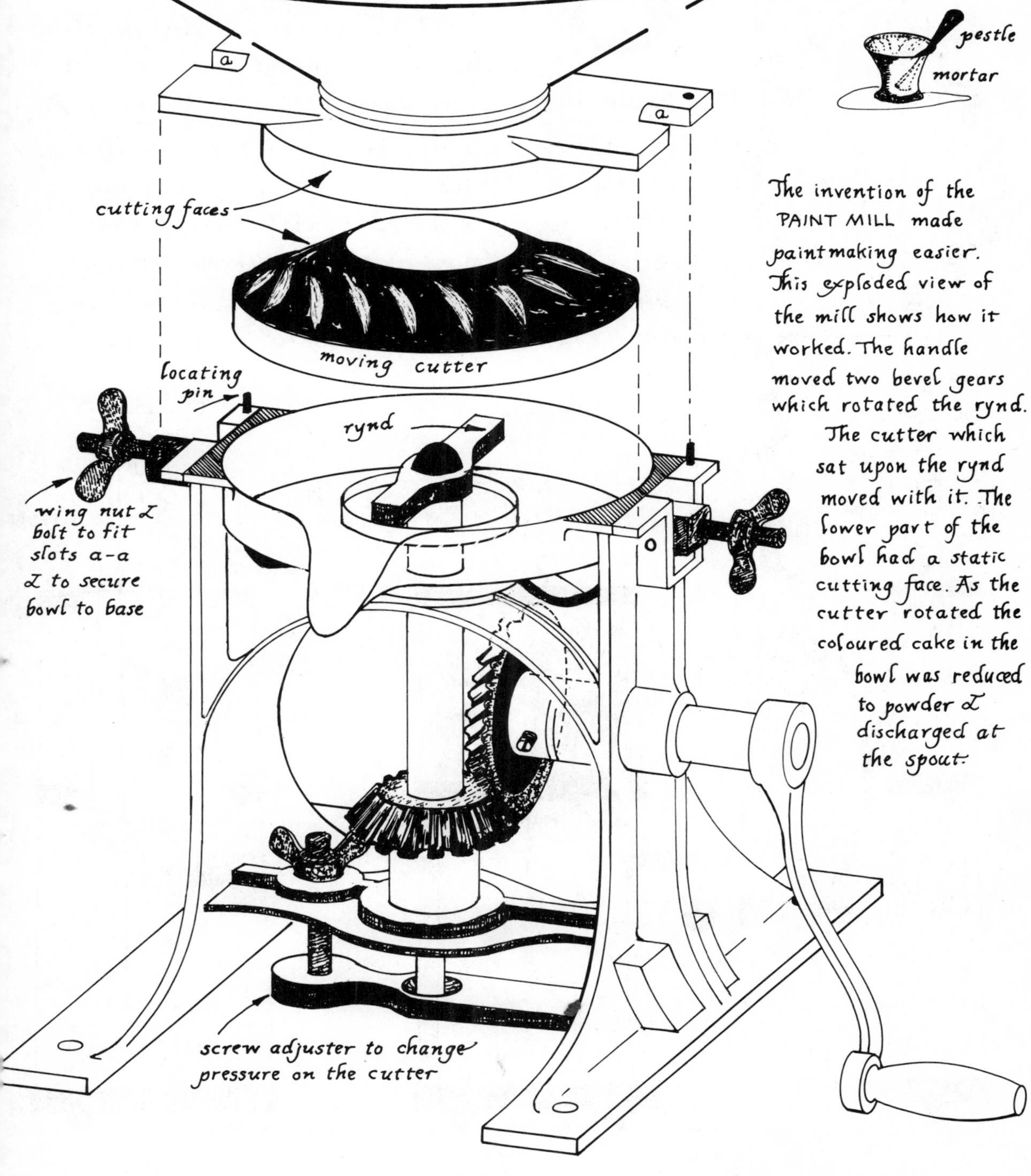

pestle

mortar

cutting faces

moving cutter

locating pin

rynd

wing nut & bolt to fit slots a-a & to secure bowl to base

The invention of the PAINT MILL made paintmaking easier. This exploded view of the mill shows how it worked. The handle moved two bevel gears which rotated the rynd.

The cutter which sat upon the rynd moved with it. The lower part of the bowl had a static cutting face. As the cutter rotated the coloured cake in the bowl was reduced to powder & discharged at the spout.

screw adjuster to change pressure on the cutter

LEONARDO'S JACK

JACKS

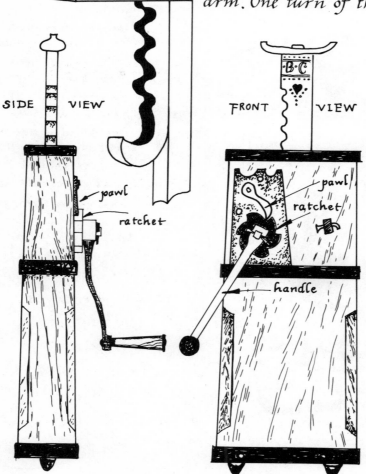

The use of gear wheels to secure a mechanical advantage was understood by Leonardo da Vinci in the C15. Jacks of the kind shown below were used as far apart as Suffolk & Conestoga, U.S.A. The handle, controlled by a ratchet, drives a four-toothed pinion (a) which operates a gear with twelve teeth. This latter gear works a three-toothed pinion which works the teeth on the jack arm. One turn of the handle moved the arm

SIDE   VIEW

pawl

ratchet

FRONT   VIEW

pawl

ratchet

handle

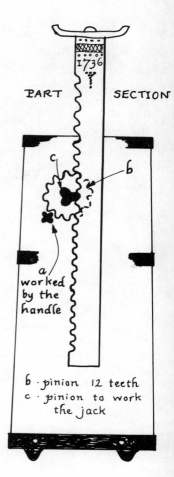

PART   SECTION

1736

c

b

a
worked
by the
handle

b · pinion 12 teeth
c · pinion to work
      the jack

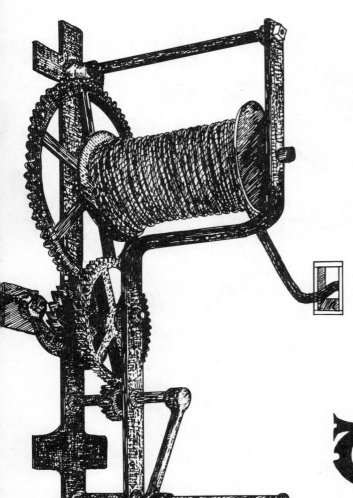

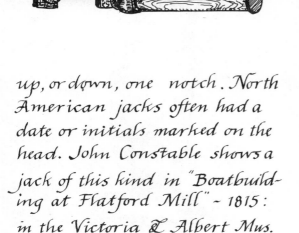

hoist

crane   pulley

END VIEW
SECTIONAL VIEW
OF BARN
WITH HOIST
& CRANE

SACK HOISTS

The development of cast iron gears was a great advance on those so laboriously cut from the wrought sheet. This example shows the manner in which a hoist could be used to lift sacks to a granary. An iron crane arm was fixed at the top of the steps. A rope from the hoist ran through a pulley at the arm's outer end. Cranes of this kind were used in many business premises. Some still remain fixed to walls near upstairs doorways.

up, or down, one notch. North American jacks often had a date or initials marked on the head. John Constable shows a jack of this kind in "Boatbuilding at Flatford Mill" ~ 1815: in the Victoria & Albert Mus.

CANAL CRANES

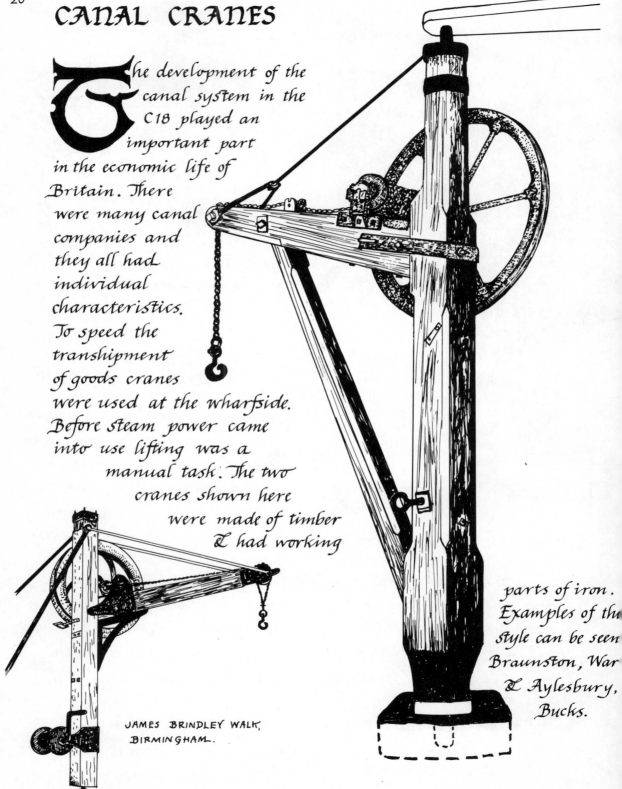

The development of the canal system in the C18 played an important part in the economic life of Britain. There were many canal companies and they all had individual characteristics. To speed the transhipment of goods cranes were used at the wharfside. Before steam power came into use lifting was a manual task. The two cranes shown here were made of timber & had working parts of iron. Examples of this style can be seen Braunston, War & Aylesbury, Bucks.

JAMES BRINDLEY WALK, BIRMINGHAM.

CRANE BARROW

WATERWAYS MUSEUM, STOKE BRUERNE.

Along the many hundreds of miles of England's waterways there was always work to be done, and very often the site was in a lonely inaccessible place. Locks needed constant care and although there is no natural current in a canal silt slowly moves downhill. Moving mud out of a lock is no easy task. One method relied upon the use of a crane barrow shown here. Its arm could be positioned clear of the lockside so that the buckets moved up and down without hindrance. When one bucket was at the top ready to be emptied the other was in the lock chamber. The winch had two handles to make the work lighter. To prevent accidents a ratchet controlled the winding mechanism.

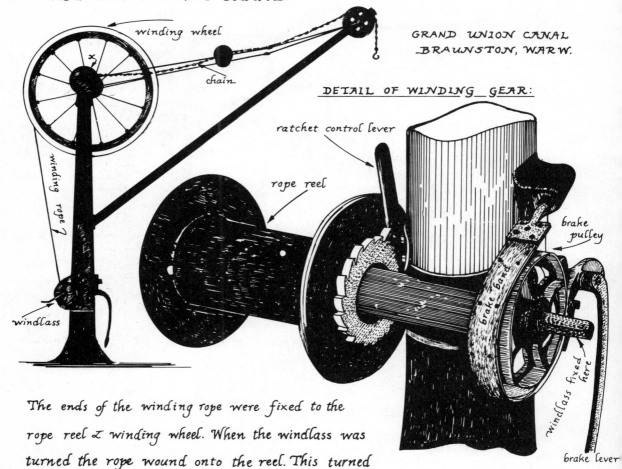

winding wheel

chain

winding rope

windlass

GRAND UNION CANAL
BRAUNSTON, WARW.

DETAIL OF WINDING GEAR:

ratchet control lever

rope reel

brake pulley

brake band

windlass fixed here

brake lever

The ends of the winding rope were fixed to the
rope reel & winding wheel. When the windlass was
turned the rope wound onto the reel. This turned
the winding wheel & the chain reel ~x. The hook was lowered by releasing the
ratchet lever. The rate of descent was controlled by the brake lever which
pulled the brake band against the pulley. Windmill brakes worked on
a similar principle ~ p.110.

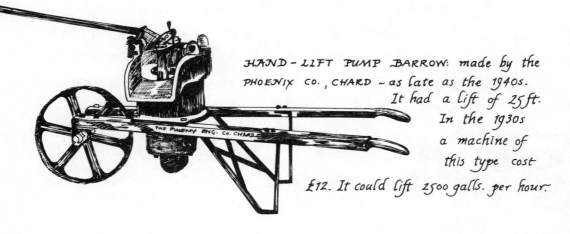

HAND - LIFT PUMP BARROW: made by the
PHOENIX CO., CHARD ~ as late as the 1940s.
It had a lift of 25 ft.
In the 1930s
a machine of
this type cost
£12. It could lift 2500 galls. per hour.

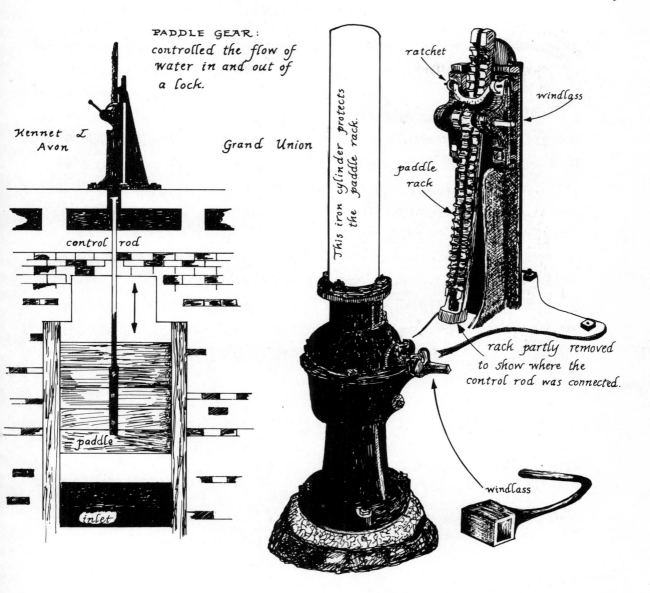

PADDLE GEAR: controlled the flow of water in and out of a lock.

Kennet & Avon

Grand Union

control rod

paddle

inlet

This iron cylinder protects the paddle rack.

ratchet

windlass

paddle rack

rack partly removed to show where the control rod was connected.

windlass

PADDLE GEAR

Paddle gear made use of a pinion wheel which worked a toothed rack. To operate a paddle a windlass must be used. The paddles shown are placed in the lock walls, but they can be in lock gates. Although paddles are heavy to lift they quickly return to the closed position once the ratchet is released.

THE HARWICH CRANE

The Ancient world made use of the treadwheel crane and the oldest documentary reference is probably to be seen in "De Architectura", circa 25 B.C., by Marcus Vitruvius. A bas relief in the Lateran Museum, Rome shows a crane operated by men standing inside a treadwheel. It seems likely that this method of lifting heavy loads remained in use during the mediaeval period, but evidence has not survived to tell us much about such machines. We will see below, pp. 28, 29 & 46, how animal power was used to work treadwheels. A crane, however, requires a degree of precision in use, and it is perhaps for this reason that lifting engines were man-powered. The crane at Harwich, Essex is a unique survivor in England. It was saved from demolition earlier this century and has been re-erected on Harwich Green. From the technical point of view it is a direct descendant of the early Roman machines.

The HARWICH CRANE has two (16ft. dia. x 3ft. 10ins. wide) wheels which share a common axle. Around the middle of this axle the lifting chain was wound. This fascinating machine is protected by a crane house, and the structure dates from the time (circa 1670) when a Naval yard had been established here. In the early C18 the yard reverted to civil use, but the crane

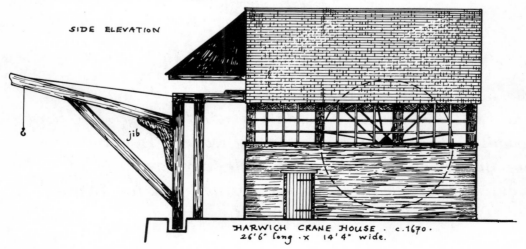

SIDE ELEVATION

jib

HARWICH CRANE HOUSE · c. 1670 ·
26'6" long · x 14'4" wide.

VIEW OF LEFT HAND WHEEL

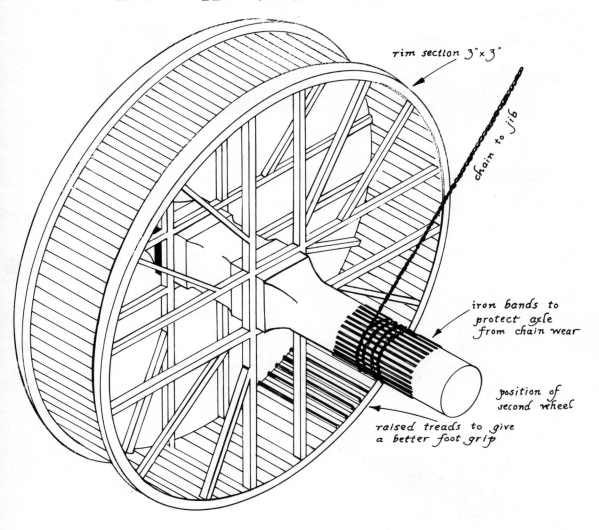

rim section 3" x 3"

chain to jib

iron bands to
protect axle
from chain wear

position of
second wheel

raised treads to give
a better foot grip

endured and was in use as late as 1914~18. Both the treadwheels
are constructed in the same way. The main spokes are arranged in
pairs to clasp the axle tree. A secondary spoke is fixed to each principal
member. There are 16 spokes on each side to support the heavy rims. The
14 ft. long axle is square in section (13½ ins. x 13½ ins) at four points to
allow the clasp arm spokes to be fixed to it with wedges. Arrangements
to view may be made on application to the Harwich Society's ~ Guide
Panel Secretary, 109 Long Meadows, Harwich, Essex.

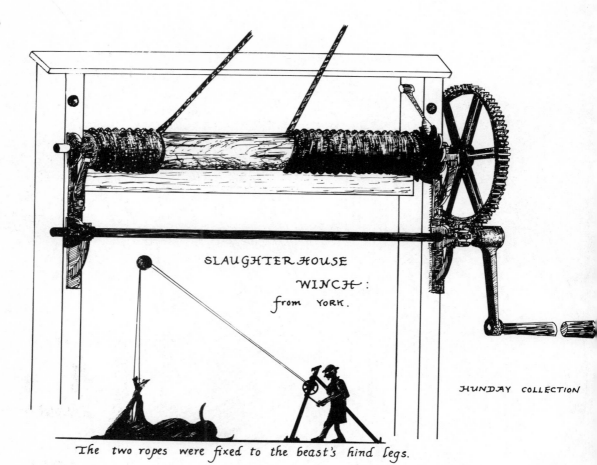

SLAUGHTERHOUSE

WINCH:

from YORK.

HUNDAY COLLECTION

The two ropes were fixed to the beast's hind legs.

HAND BELLOWS: Every blacksmith needed a pair of bellows at his forge. Keeping the fire hot was often a task for a boy. Industrial hearths sometimes had a mechanical alternative to this ancient device ~ see p. 84.

ARMLEY MUSEUM

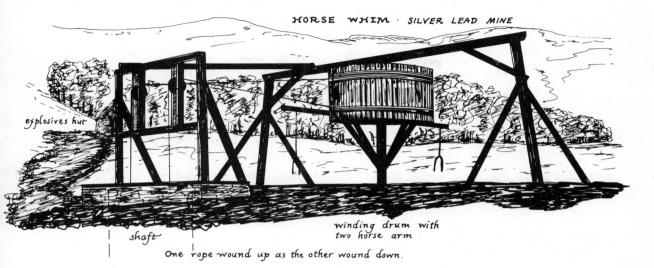

HORSE WHIM · SILVER LEAD MINE

explosives hut

winding drum with
two horse arm

shaft

One rope wound up as the other wound down.

ANIMAL

For centuries on end the work of the quadruped was limited to carting, ploughing and threshing with its feet. It was the advances man made in technology during the early modern period that opened up a new role for animal power. Perhaps the inspiration was derived from the mediaeval dog-operated spits found in castle kitchens. From the C17 onwards engineers began to devise machines that could be worked by animal power. Treadwheels were once widely used in England for raising water from wells. Carpenters made horseworks like the example of the buttermill in the Science Museum. Cast iron made the construction of smaller machines possible, and these were used in industrial & agricultural settings. Cattle, donkeys and dogs were also put to work at various tasks.

~ DONKEY WHEELS ~

WEALD & DOWNLAND MUSEUM·
CATHERINGTON WELL HOUSE.

A water supply often decided the site of a human settlement, and several thousand English villages bear witness to this basic need. For hundreds of years water was lifted from wells with a windlass, or a counter-balance like the one shown on page 142. Mechanical methods of water raising, in England, probably began in the C16. The well-known donkey wheel at Carisbrooke Castle, I.O.W., dates from 1687. Another wheel of note can be seen at Greys Court, Oxon. ~ National Trust. Its 19ft. wheel worked two buckets which were automatically tipped up at the top of their journey. We usually call them donkey wheels although they could be worked by manpower. The Upham wheel opposite with its 2ft. wide treadwheel was intended to be worked by two legs. These early machines were made by carpenters and they possess a minimum of ironwork ~ except where later repairs required it to add strength to worn timbers. Examples of water raising wheels have been recorded as far apart as Devon and Yorkshire ~ see p.156. These machines represent an important stage in our technical history as they were all individually made, and show how different craftsmen solved the same problem.

The Beech wheel is 3ft. 8ins. wide & 12ft. in diameter. Its four main spokes are half-lapped & wedged, each with two braces. Between the two rims are the planks which make the tread. Winding up and lowering required the operator to face in alternative directions.

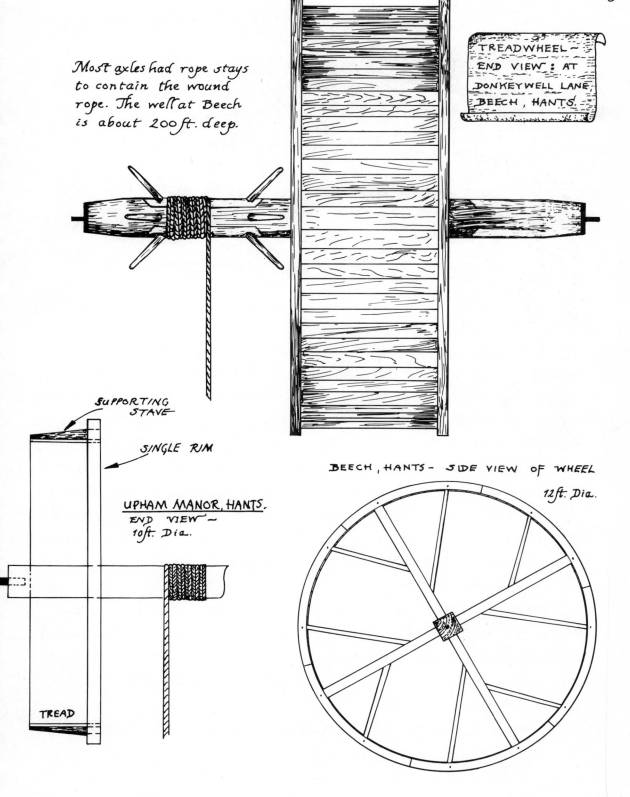

Most axles had rope stays to contain the wound rope. The well at Beech is about 200 ft. deep.

TREADWHEEL ~ END VIEW: AT DONKEYWELL LANE, BEECH, HANTS.

SUPPORTING STAVE

SINGLE RIM

UPHAM MANOR, HANTS. END VIEW ~ 10 ft. Dia.

TREAD

BEECH, HANTS ~ SIDE VIEW OF WHEEL

12 ft. Dia.

DONKEY POWER

koggelstock

disselboom

onkey power was also used in irrigation schemes in hotter parts of the world. Small waterwheels, with iron buckets to lift water from one level to another, were manufactured in various places. The example drawn here shows a typical South African donkey-powered wheel which was very similar to the horse gear on page 39. A long wooden arm was attached to the wheel gear and the donkey was harnessed to the swingle tree at the end of the arm. To keep the donkey walking in a constant path a cane ~ a kogglestock ~ was also fixed near the inner end of the arm, and to the donkey's bridle.

Such a simple device could be made by the village blacksmith and there were a wide variety of design differences to be observed. Many wheels were constructed from spare parts salvaged from other machines. Gear wheels were seldom thrown away as they could always be re-used.

Power from the donkey was transferred to the horse-arm by way of a simple breast strap. The ends of the breast strap were hooked to the trace chains that were in turn attached to the swingle hooks. Harness of this kind was used in Antiquity and can still be seen today in southern Italy.

The circular path trodden by the donkey was transformed into a vertical motion by the use of bevel gears. One disadvantage of these shallow wheels was the small amount of water raised at each revolution. All that was needed was a boy to keep the animal moving.

Bucket-pumps of this kind were called 'bakkiespomps' in Afrikaans. They lifted water in the same way as the noria shown on p.47. James Walton - p.157 - has shown that this method of water raising was very common in the homesteads of Transvaal in the 1900s. Donkey or mule power was also applied to snuff grinding, which was an important commodity in Cape Town in the latter part of the C19.

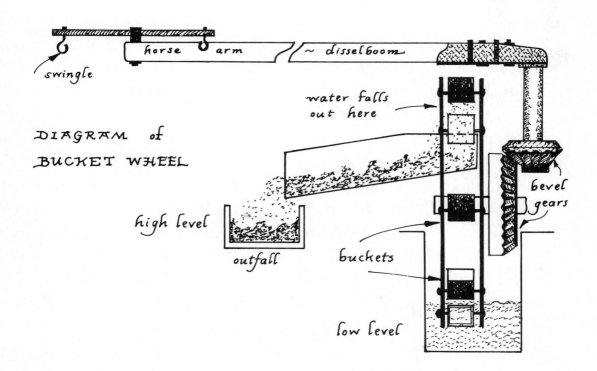

DIAGRAM of BUCKET WHEEL

swingle

horse arm ~ disselboom

water falls out here

bevel gears

high level

outfall

buckets

low level

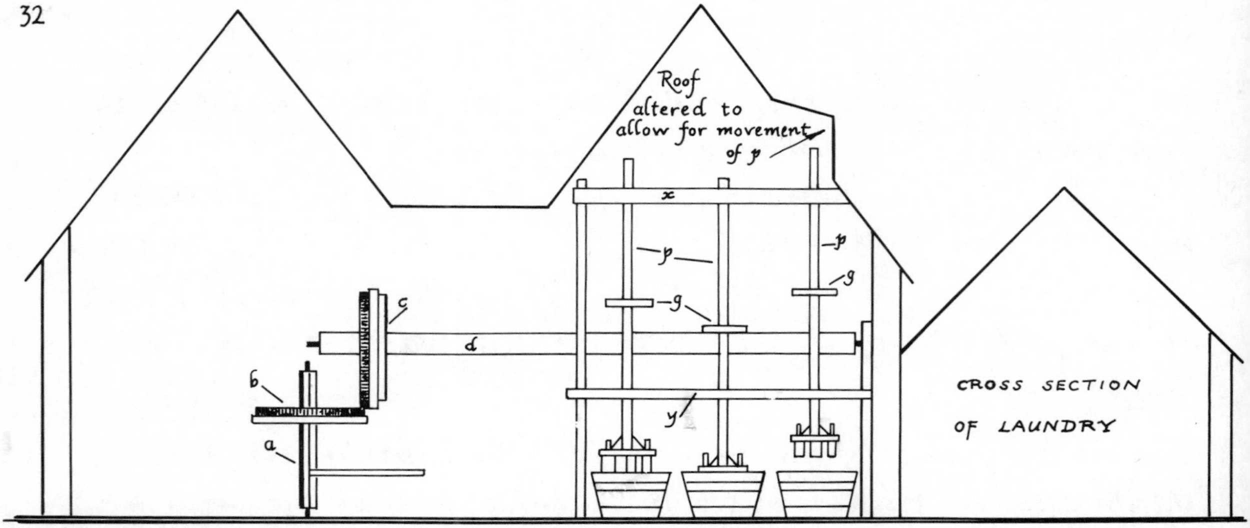

Roof altered to allow for movement of *y*

x

p

p

g

g

c

d

b

a

y

CROSS SECTION OF LAUNDRY

THE HORSE-POWERED LAUNDRY

Both the cottager and the laundry maid in the great house were familiar with the dolly that was pounded up and down in the dolly tub to release dirt from the wash. In Holland mechanics came to the bleaching trade, where a similar agitating motion was needed, as early as the C17. A horse engine was evolved that made it possible for one horse to operate several large agitators at the same time. One such fascinating engine has been preserved at Arnhem – see p. 154. Originally it was used in the linen bleaching trade, but as that industry declined in the C19 many bleacheries became launderies. In the C18 wealthy families sent their linen to the bleachery once or twice a year.

After boiling, soaking and rinsing linen was pounded by the beaters in baths of warm suds (k). The diagram above shows: (a) the vertical shaft and gear (b) moved by the horse; the secondary gear (c) which turns the layshaft (d). As (d) rotates its four cams (f) lift in turn one of the pounders (p). The lift comes from a cam pushing against a lifting plate (g). The cams are placed on different faces of (d) so they operate in sequence. Each pounder is lifted four times by one rotation of (d). As each cam lifts (g) it also rotates it about 90°. Thus the position of each pounder changes at every lift. This engine anticipated many of the features we now expect to find in a modern washing machine.

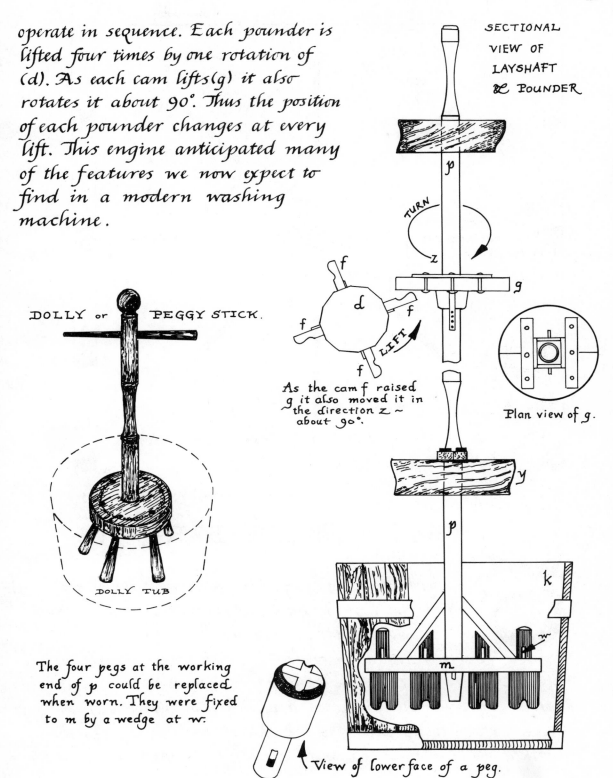

SECTIONAL VIEW OF LAYSHAFT & POUNDER

TURN

z

g

As the cam f raised g it also moved it in the direction z ~ about 90°.

Plan view of g.

DOLLY or PEGGY STICK.

DOLLY TUB

y

p

k

w

m

The four pegs at the working end of p could be replaced when worn. They were fixed to m by a wedge at w.

View of lower face of a peg.

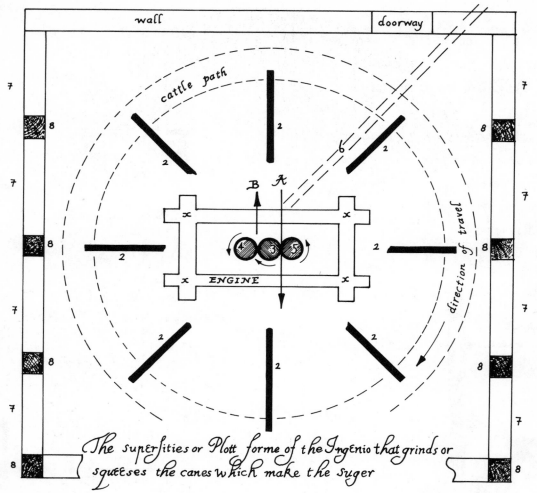

The superfities or Plott forme of the Ingenio that grinds or squeeses the canes which make the suger

PLAN OF CATTLE MILL FROM RICHARD LIGON'S "HISTORY of BARBADOS" · 1657.

CATTLE MILLS ~ BARBADOS

The manufacture of sugar in Barbados dates from 1641. In the first mills built to extract the juices from sugar cane cattle were used as the source of power. Richard Ligon provides us with considerable detail about these early engines. From a vertical shaft set at the centre of the mill projected a number of sweeps (2). The cattle were attached by chain to the end of each sweep. As they walked round their set path the shaft turned the middle

Cattle mill with open sides. Juice from the mill flowed downhill to the boiling house.

roller (3). In their turn rollers (4 & 5) moved in the way shown by the arrows. Cane was passed through the rollers twice. First it was crushed between the rollers shown by arrow A. Then a second worker passed it back ~ arrow B ~ through the engine. Juice extracted flowed into a channel (6) beneath the floor, and into a cistern beyond the wall.

To work well in such a hot climate cattle need adequate shade and good ventilation. The mill's high thatched roof gave shelter from the sun, and its open sides (7) divided by posts (8) allowed the breeze to pass through the mill. A heavy tie beam held the upper end of the engine shaft in place.

In 1683 there were 358 sugar works in Barbados, but by that time windpower had been introduced ~ p. 132. During the last decade of the C17 cattle & windmills existed in equal numbers. Twenty years later the pattern had changed. Some 76 cattle mills then survived among nearly 500 windmills. From the middle of the C19 steampower was used. The more efficient steam engine eventually replaced both cattle and windpower.

This cart is a hermaphrodite ~ see 'OLD FARMS' p. 75.

Cattle power was also used to carry cane from the fields.

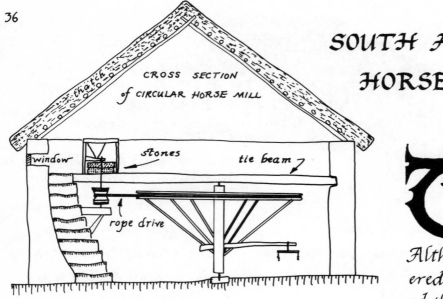

CROSS SECTION
of CIRCULAR HORSE MILL

window

stones

tie beam

rope drive

SOUTH AFRICAN HORSE MILLS

The first corn mill in South Africa was built in 1657. Although water-powered mills soon followed the horse mill remained a useful machine in areas where streams were not reliable. Some of the horse mills had a simple rope drive from the horse-wheel to the pulley that turned the spindle shared by the runner stone. In the diagram below an under-driven stone is shown. Some horse mills had over-driven stones. In addition to the rope-driven stones many mills had a cogged horse wheel which drove a lantern wheel as the drawing on the right indicates. The vertical shaft was made from a single timber about one foot in diameter and some nine feet long. From the horse's point of

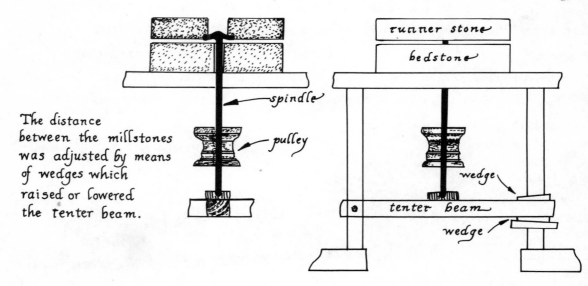

The distance between the millstones was adjusted by means of wedges which raised or lowered the tenter beam.

spindle

pulley

runner stone

bedstone

wedge

tenter beam

wedge

view milling was a tedious task and to make it bearable the animal was blinkered. To make the horse move in a regular path a cane, a koggelstock, was attached to the shaft and to the horse's bridle. This kept the horse at a fixed distance from the wheel's centre.

The arm, a disselboom, was subject to a considerable force. Where it was fixed to the vertical shaft it was strengthened by iron straps. The cogged mills were more effective than the rope-driven type as their drive was more positive.

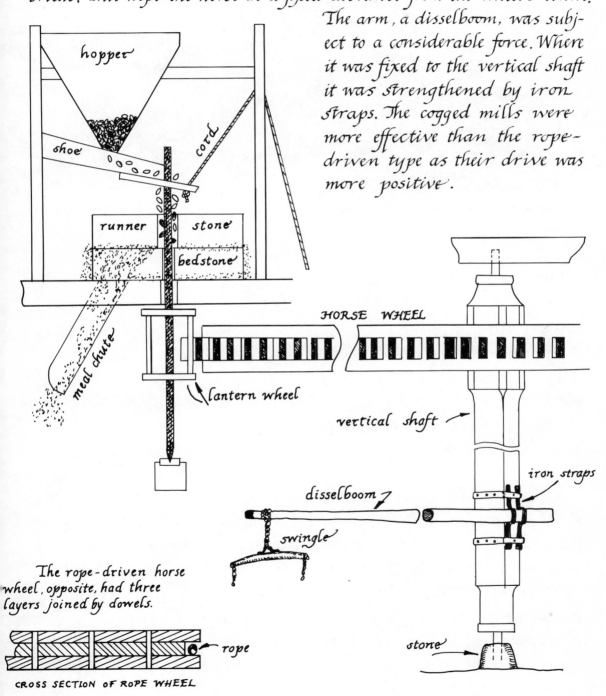

hopper

shoe

cord

runner stone

bedstone

meal chute

lantern wheel

HORSE WHEEL

vertical shaft

iron straps

disselboom

swingle

The rope-driven horse wheel, opposite, had three layers joined by dowels.

rope

stone

CROSS SECTION OF ROPE WHEEL

WAGON UNLOADING   ELEVATOR   DRIVING RODS   HORSE GEAR

HORSE GEAR

In the C19 farmers were quick to take advantage of the many new horse-operated machines which engineers devised to reduce the work of the farm labourer and save his master money. One of the most useful inventions was the horse gear which required the horse to walk round a set of gears. The power generated was taken off the gears by a series of driving rods linked together by universal joints. On some farms a shelter, usually circular in plan, was built to protect the machinery & the horse. Then the gear was linked to machines inside adjacent barns. A great advantage of horse gear was its small size which made it portable. In Victorian times horse gear found its way into the harvest field to operate elevators which made the building of corn or hay stacks quicker and therefore cheaper. Portable horse gear was towed behind the elevator on its own small trailer. Horse gear could be operated by a single horse,

TWO-HORSE GEAR

horse arm

horse arm

universal joint

spur wheel

crown wheel

but some tasks required two, three or even four horses – one on each arm. The speed of a horse around a gear set was about two miles per hour. At this speed a horse provided some 160 lbs. of tractive effort. If the speed increased the effort diminished. The prudent farmer's interests were best served by the plodding horse. In the 1900s a farm horse's keep cost its owner half a crown (12½p) a day. A farm worker's wages were 3½d (2p) an hour! Horse gear was made from the early years of Victoria's reign. An 1847 catalogue listed a Birmingham manufacturer's apparatus (Ryland & Dean) at the following prices: one or two horse gear £14-0-0; one to three horse gear £17-0-0; one to four horse gear £22-0-0. The technical excellence of horse gear was founded on the universal joint which allowed it to be operated at variable angles on uneven ground.

The idea of placing a millstone on edge and running it in a circular path on a flat stone bed is very old. Mills of this kind were put to many uses and so robust were their crushing pans that many still survive long after the millstone and its fittings have gone. England's western counties were once renowned for their cider orchards. There was a time when cider mills like the one shown above were to be seen in almost every village. Particularly good examples preserved in their working state can be seen at Mary Arden's House, Wilmcote; Hartlebury Castle, Kidder-

minster; & the Museum of Cider, Hereford. Examples of these old mills can be seen in country gardens or as roadside ornaments and they have also given their name to a number of inns in midland counties.

The antiquity of this style of mill is illustrated by the Egyptian olive press above which could be operated by mule or man. Like the other presses it has a central post that turns 360°. The axle is attached to this upright post, and extends beyond the crushing pan. This long lever is pushed or pulled in order to move the stone around the upright shaft.

Edge runner stones, with a weight of a ton or more, were used by C18 industrialists to crush materials such as flint and bone. These materials were employed in pottery and glass manufacture. Wind and water power was also applied to edge mills.

Ye Olde Crab Mill

OIL MILLS

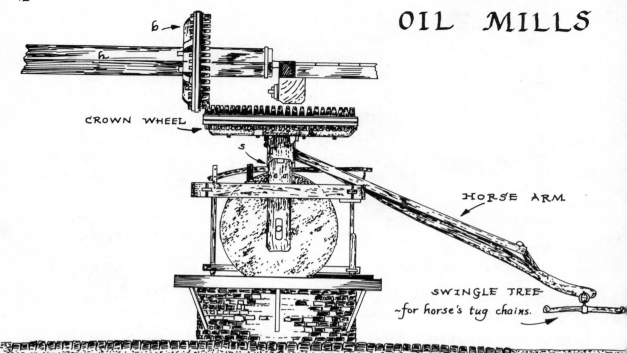

CROWN WHEEL

HORSE ARM

SWINGLE TREE
~for horse's tug chains.

By the C19 men had become practised in the methods of extra-cting oil from the seeds of flax - linseed - rape, cole and many other plants. One of the most impressive horse mills is the Dutch one shown here. It has two stones, 57½ ins (146 cms) diameter. As the horse walked around the crushing pan the stones moved about the central shaft. The large crown wheel ~ 48 teeth - turned at the same rate. Its cogs moved the vertical gear (b) ~ 32 teeth ~ and this turned the heavy horizontal shaft (h). At the opposite end of the shaft there was another smaller gear which operated the press ~ page 44. In this way the momentum of the moving stones helped to drive the ancillary machinery. The stones of this edge - mill, called a KOLLERGANG in Holland, are not equi-distant from the upright shaft (s). They therefore roll a different path around the crushing pan. A wooden arm with a curved striker at its outer end moves with

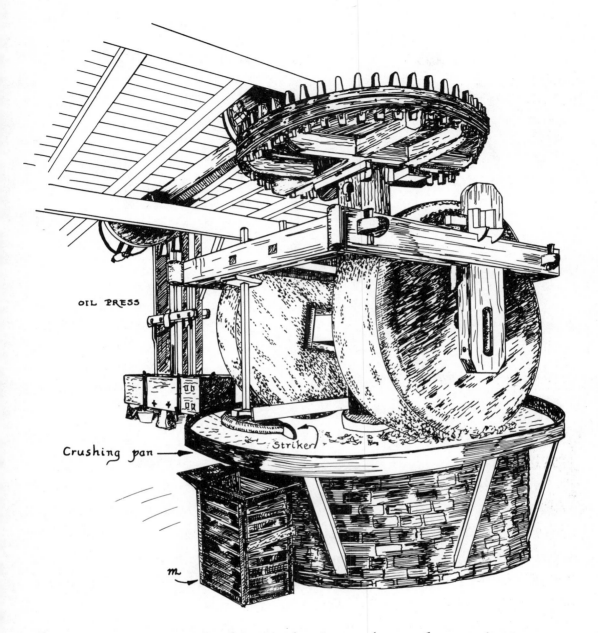

OIL PRESS

Crushing pan ⟶ Striker

m ⟶

the stones and distributes the linseed, or other seeds, across the surface. The action of the stones reduces the seed to a pulpy mash. A sliding lid in the parapet can be removed to allow the 'seed meal' to be scooped into the meal box (m). It can then be taken to the pan to be heated before it is placed in the press.

The workhorse was changed every two hours.

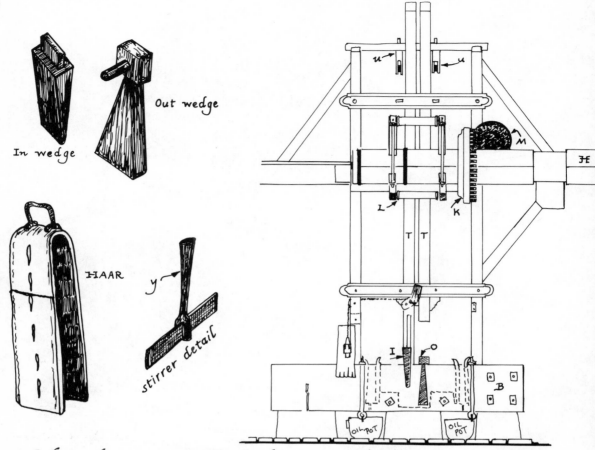

In wedge

Out wedge

HAAR

stirrer detail

Before oil is extracted from the seed-meal it has to be heated in a pan (P). This pan has a mechanical stirrer (Y) that is moved by the following train of gears. The wheel (K) is on the horizontal shaft moved by the kollergang; (K) moves (M) & (N); which turns (Q). The stirrer (Y) is at the bottom of shaft (R). It can be disengaged by pulling rope (S) that raises (R) out of gear.

The OIL PRESS is also operated by the kollergang's motion. As (H) rotates it moves the gear (K) and the lifting wheel (L). This latter wheel can lift the two rams (T) so that they fall and strike the 'out' or 'in' wedges in the press below. The rams can be disconnected from their lifting wheel (L) by the pulleys and levers (u & v).

The body of the press (B) is a massive timber almost 24 ins. square. At its centre there is a cavity. In this space the woollen bags of meal were placed wrapped in their horse hair cases - haren - with the

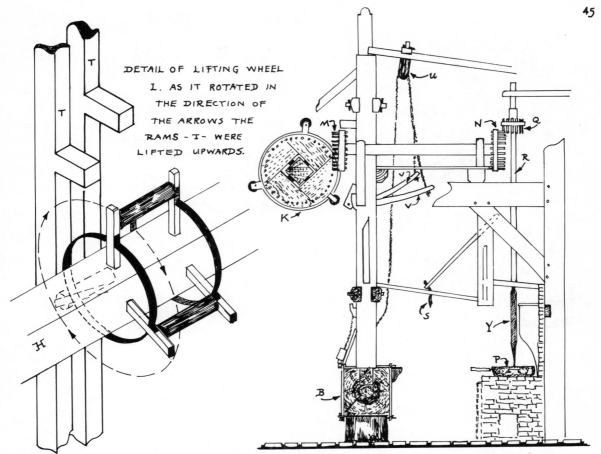

DETAIL OF LIFTING WHEEL
L. AS IT ROTATED IN
THE DIRECTION OF
THE ARROWS THE
RAMS -I- WERE
LIFTED UPWARDS.

wedges. The 'in' wedge (I) was struck several times by its ram. This action squeezed the contents and the oil was discharged. It flowed into one of the metal containers below the press. When the 'out' wedge (O) was struck it released the pressure and the bags could be taken out. The meal, by pressure, became a solid cake. This was separated from its haar on the upright member (W) of the cutting bench ~ koekenhok. The cake could be processed again to produce a lesser quality of oil. Before the chopped cake was placed on the koller~ gang it was made smaller in the hand operated ~koekbreker.

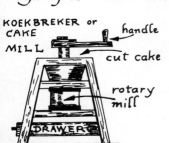

KOEKBREKER or
CAKE
MILL

handle

cut cake

rotary mill

DRAWER

A press of this kind could produce about 10 gallons ~40 litres~ of oil each day, in ten or twelve hours.

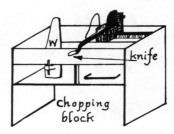

W

knife

chopping block

DUTCH DOG CHURNS

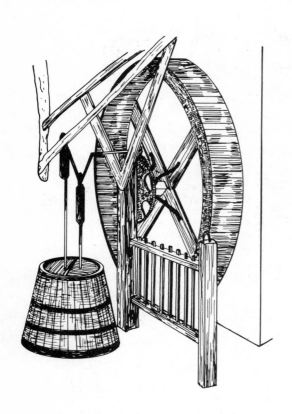

Making butter by hand was always time consuming. Dutch farmers, or more likely their wives or daughters, made use of dog power to work the churns. This dog wheel next to a wall has an open side partly blocked by a fence. The staple on the post was to secure the dog's lead. As the wheel turned it rotated the spur gear, and this moved a smaller pinion - not visible - on the crank axle. The two plungers then moved up and down in the churn.

The example below also has a cranked axle with two rocker arms that operate the plungers. Motion was generated by a dog inside the sloping pen. As the dog walked forwards the slatted floor ~ a continuous belt ~ also moved. This action turned the drum x which is fixed to the crank axle.

Both these machines can be seen at the Netherlands Open Air Museum, Arnhem.

WATER

The oldest representation of man's use of water power is probably the mosaic of a waterwheel at the Great Palace, Byzantium. Examples of the water raising NORIA, which dates from Antiquity, can still be seen in the Middle East. Thousands of mills are recorded in William of Normandy's Domesday Book ~ 1086. Many extant mills claim an origin in that early period. They may occupy the site of an ancient mill but their fabric will have been replaced many times. Domesday records such as ' 2 mills at 6s. 8d.' do not suggest that early mills were very large. When water power was applied to early industrial processes in the C18 some very large mills were built – like the one at Quarry Bank, Styal, Cheshire. By Victoria's reign millwrights had developed substantial corn mills too. Some like this one had more than one waterwheel and so had a high capacity output.

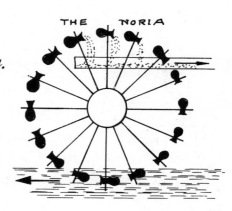

THE NORIA

HORSTEAD MILL, COLTISHALL, NORFOLK.

THE NORSE WATERMILL

The history books may tell us that the feudal system was replaced by the C18 but in many places it was still present a century later. Early in the C19 the Isle of Lewis was owned by one man. He was able to insist, via his factors-agents, that crofters' grain should be ground at his own mills. The toll for grinding was one sixteenth part of the grain. A Royal Commission investigating conditions in the Highlands and Islands ~ 1883 ~ heard from Norman MacDonald, of Carloway, that crofters were prohibited from using their own mills. This did not mean that private mills were not used. Away from the road and hidden by the hills many were worked. The old Mill of the Blacksmiths ~ Muilinn Nan Gobhaichean ~ is located near Loch Raoineabhat at Shawbost (reconstructed 1970). Shared by all the villagers it derived its name from the blacksmith brothers who built it.

The NORSE MILL with its horizontal wheel was used by the Viking colonizers in the C8. It has a very simple mechanism. Water was directed onto one side of the vanes of a vertical tirl. At its upper end the tirl passes through the un-moving bedstone to support and propel the upper runner stone. Grain ran from the hopper onto the feed shoe that was agitated by the curved wooden clapper. Tied to the feedshoe its lower end ran on the upper face of the granite millstone. The vibration caused the oats or barley to fall into the eye of the stone.

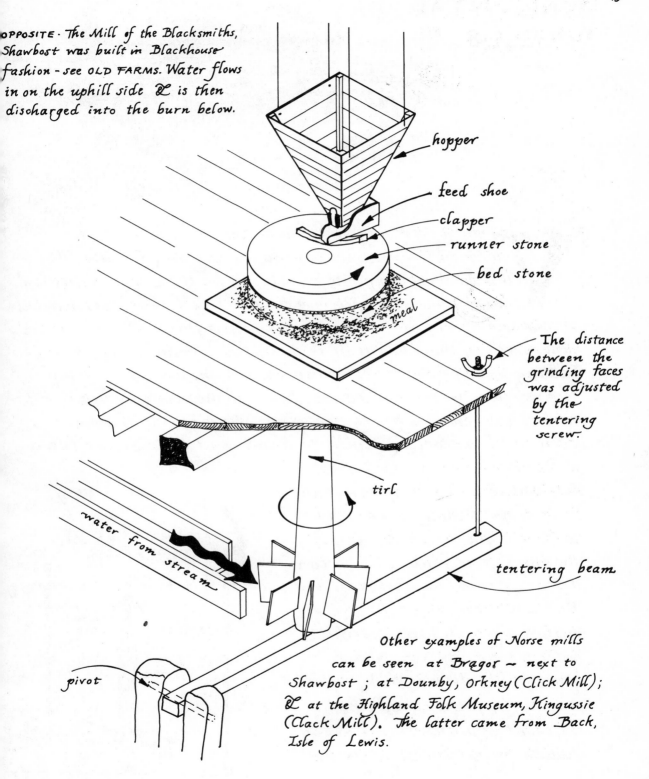

OPPOSITE · The Mill of the Blacksmiths,
Shawbost was built in Blackhouse
fashion – see OLD FARMS. Water flows
in on the uphill side & is then
discharged into the burn below.

hopper

feed shoe
clapper
runner stone
bed stone

meal

The distance
between the
grinding faces
was adjusted
by the
tentering
screw.

tirl

water from stream

tentering beam

pivot

Other examples of Norse mills
can be seen at Bragor – next to
Shawbost ; at Dounby, Orkney (Click Mill);
& at the Highland Folk Museum, Kingussie
(Clack Mill). The latter came from Back,
Isle of Lewis.

HORIZONTAL WHEELS

CLICK MILL, DOUNBY ~ ORKNEY.

Vikings certainly introduced the technology of the horizontal wheel to Iceland ~ where small mills like the one shown opposite were in use in the C19 ~ a millenium later. The distribution of the horizontal wheel has puzzled archaeologists, but it certainly was not confined to areas directly influenced by the Vikings. It is the author's opinion that a long-lasting legacy from the Norsemen caused the horizontal wheel to find its way to South Africa. Most of the South African mills were small ~ about 12ft. x 15ft. in plan. The farmers who built them often used materials ready to hand, and they do not follow a standard design. Like most horizontal mills they were built on a slope so that it was easier to direct the water supply onto the wheel. Horizontal mills can also be found in Tuscany and this diagram shows the drive system used. All its essential features are shared by the other mills discussed. From the outside the Tuscan mills have a different appearance reflecting Italian architectural ideas, and

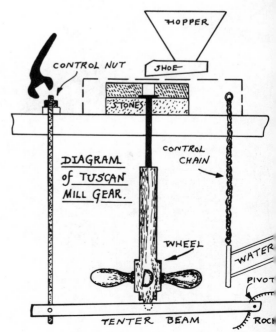

HOPPER

CONTROL NUT

SHOE

STONES

DIAGRAM of TUSCAN MILL GEAR.

CONTROL CHAIN

WHEEL

WATER

PIVOT

TENTER BEAM

ROCK

SOUTH AFRICAN MILLS

GAMAS KLOOF ~ LADYSMITH

HOEKO ~ LADYSMITH   OUTFALL

the fact they often have three or
four pairs of stones. This makes a
distinct contrast to the Icelandic
mill which is hardly as tall as a
man.

There are two styles of horizontal
wheel. The Norse wheel has a heavier
tirl and fewer blades than the
Mediterranean type with its
slender drive shaft and shallow
spoon shaped blades. Horizontal
mills are also to be found in Spain,
Greece, Roumania, Israel and
Lebanon.

FOUR STONE MILL ~ LIMA VALLEY

C19.
ICELANDIC MILL

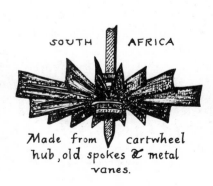

SOUTH AFRICA

Made from cartwheel
hub, old spokes & metal
vanes.

NORWAY

VERTICAL WATERWHEELS

The waterwheel can be described by its manner of rotation or by its structure. There are several ways of driving a wheel. The simplest, and most inefficient, method of working was to use a paddle or dipper wheel which ran in a stream without an effective head of water. A better result was achieved by raising the water delivery point so that it was near or above the level of the wheel's axle. This produced more power. Wheels of this kind are called undershot or breastshot wheels. If the delivery point was level with the wheel's centre the wheel was also termed a 3 o'clock wheel. Mills could have 2 o'clock or 4 o'clock wheels which were also called high or low breast wheels. The former type produced more power than the latter as the water travelled further round its circumference before it was discharged into the tail race. During the C18 several engineers including Smeaton began to study the action of water upon a wheel's surface, and calculations for the relative proportions of a waterwheel started to be made. Jean-Victor Poncelet (1788 - 1867) made the greatest improvement in the design of the undershot wheel. He introduced curved paddles and a similar profile to the race. The most efficient wheel was overshot. Its water affected the wheel's rotation for the greatest distance. The pitch back or back shot wheel was almost as effective. A good working example of the latter can be seen at Llywernog Silver-Lead Mine, Dyfed.

Early waterwheels were wooden, with little ironwork. Axles on wooden wheels had to be large. To avoid weakness the spokes were not mortised into the axle but arranged in clasp arm fashion with each spoke touching one side of the square axle tree. Adjustments to wheels of this kind were made by wedges between the axle and spokes. As the use

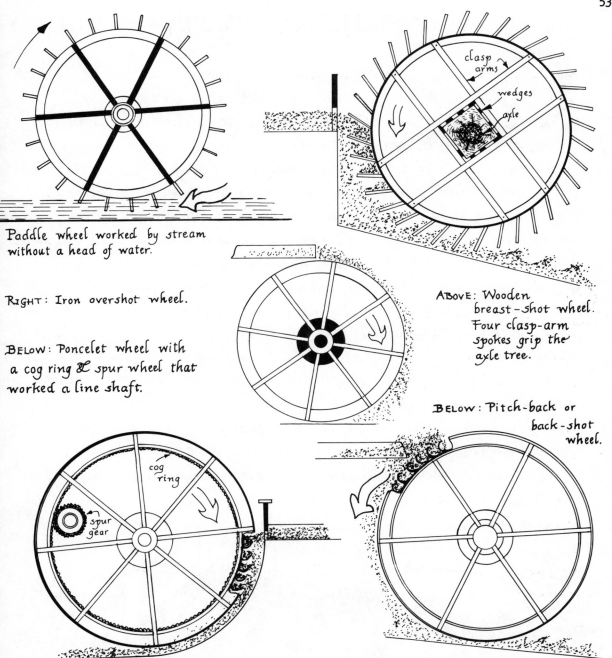

Paddle wheel worked by stream without a head of water.

RIGHT: Iron overshot wheel.

BELOW: Poncelet wheel with a cog ring & spur wheel that worked a line shaft.

clasp arms

wedges
axle

ABOVE: Wooden breast-shot wheel. Four clasp-arm spokes grip the axle tree.

BELOW: Pitch-back or back-shot wheel.

cog ring

spur gear

of cast iron developed in the C18 engineers began to make iron wheel centres which ran on forged iron axles. Eventually wheels were made entirely of wrought & cast iron. Some wheels had cast iron centres & rims, but wooden spokes & paddles.

WATERCOURSES

Control of the water supply was a basic need for the miller. Water passed through the wheel race when the mill was working. At other times it had to be stored in a mill pond or, if the mill was on a river, discharged via a sluice or a weir. The location of a mill determined the type of wheel used. In a hilly area water could be diverted from a stream when needed and allowed to follow its natural course at other times. Many mills in upland areas were built below the level of the watercourse, & they often had overshot wheels. To bring natural water to the point where its power was required sometimes made it necessary to dig a long leat or, in hilly places, a wooden launder or lade ~ a trough supported on trestle legs. In lowland places a natural stream was often dammed so that a large amount of water could be held back for the miller's use. To avoid a flood a dam had to be provided with a by-pass of some kind so that excess water could drain away. A dam also provided a bridge across the watercourse. Mill bridges often became busy thoroughfares and, particularly in towns, had an influence on street plans.

LAUNDER, FINCH FOUNDRY, STICKLEPATH, DEVON.

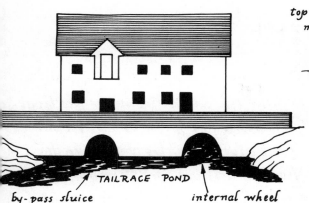

Mill built below the dam.
Notice the shallow arches.

by-pass sluice   TAILRACE POND   internal wheel

A mill built above the dam & upon the bridge.
The tailrace was a wide pond which could
take a large amount of water from upstream
without flooding. It also served as a water
supply for farm stock & the miller's horse.

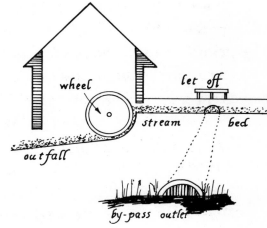

wheel   let off

stream   bed

outfall

by-pass outlet

Cross-section of mill built below a
dam with subterranean by-pass
controlled by a sluice or let-off.
The by-pass emerges at the foot of
the dam and rejoins the main
stream below   the mill.

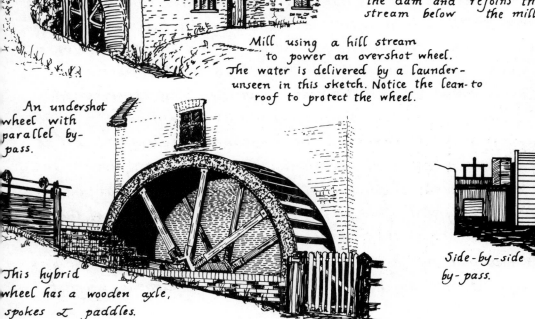

Mill using a hill stream
to power an overshot wheel.
The water is delivered by a launder-
unseen in this sketch. Notice the lean-to
roof to protect the wheel.

An undershot
wheel with
parallel by-
pass.

Side-by-side
by-pass.

This hybrid
wheel has a wooden axle,
spokes & paddles.

DRIVING THE STONES

The train of gears devised to take the drive from the vertical water wheel remained unchanged for more than a thousand years. This fact alone speaks volumes for the insight and skill shown by those first engineers who translated natural power into both vertical & horizontal planes. From the days of Antiquity the millwrights' basic material was wood, later augmented with a modest amount of wrought iron to strengthen moving parts where mechanical stresses were greatest. Old wooden machinery was heavy, & the volume of timber required to provide sufficient strength limited the size of the machine man could contrive. The advent of cast iron in the C18 gave engineers a new freedom to apply to their machines. Cast iron gears with wooden teeth running together with wholly iron wheels made for greater precision and smoother operation.

This diagram shows how the vertical wheel's power was finally made to turn the horizontal millstones. The wheel A is mounted on the axle B-which is shared by the pit wheel C. As they both turn together the pit wheel turns the wallower E. This is mounted on the vertical shaft S; also shared by the great spur wheel F; and the crown wheel Q. The spur wheel turns each stone nut G which, via a stone spindle P, causes the runner stone I to rotate. The bedstone H does not turn at all. Each pair of stones is protected by a wooden case, tun or vat J. Upon the vat rests the 'horse' M which supports the hopper L containing the grain. This is fed by gravity into the eye of the runner stone N; as the feed shoe is shaken by the damsel T or quant – see page 114. Before the grain can reach the hopper it has to be carried aloft to the corn bins on the top floor by way of

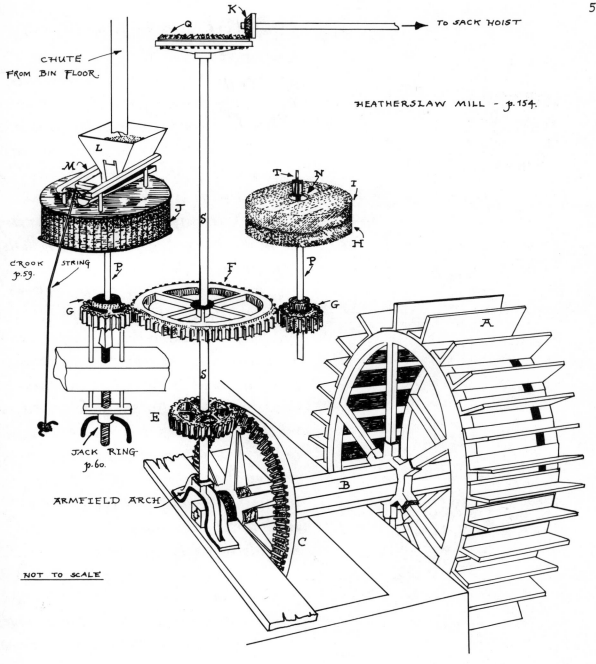

TO SACK HOIST

K

Q

CHUTE
FROM BIN FLOOR.

HEATHERSLAW MILL - p.154.

L

M

T N

I

J

S

H

CROOK
p.59.

STRING

P

F

P

G

G

S

JACK RING
p.60.

E

ARMFIELD ARCH

A

B

C

NOT TO SCALE

the sack hoist operated from the crown wheel Q and the bevel
gear K. Sack hoists are illustrated on pag 66.

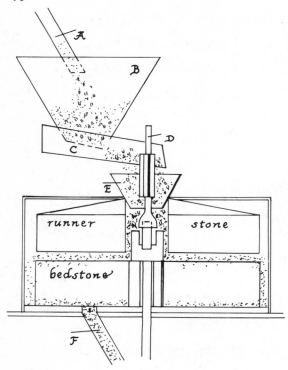

runner   stone

bedstone

B

H

C

crook string   VAT or TUN

GRINDING the GRAIN

To those unfamiliar with milling a first visit to a working mill is something of a mystery. They see and hear the motion of the machinery and watch the grain fall into the eye of the stone helped by the chattering damsel. It is not so easy to see whence the grain came and where the flour goes. Many centuries before Newton discovered the laws of gravity millers were using it to feed grain to the stones, and to tumble meal into the ark. This diagram shows the path followed by the grain. Gravity allowed the grain to descend from the bins, on the top floor, via chute A to the hopper B. From there it trickled into the shoe or slipper C which was agitated by the damsel D. The grain then fell into the stone's eye E where it came into contact with the cutting edges ~ see p.64. Both stones were cut in the same manner. When they were placed face to face the grooves were opposed. As the runner stone moved above the stationary bed stone the

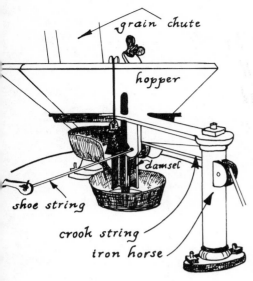

grain chute

hopper

damsel

shoe string

crook string

iron horse

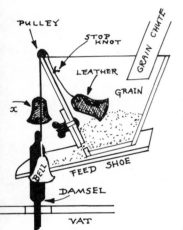

PULLEY

STOP KNOT

LEATHER

GRAIN CHUTE

GRAIN

x

BELL

FEED SHOE

DAMSEL

VAT

edges of the furrows passed across one another in a scissor-like action. The ground grain was expelled around the skirt where it fell into the meal chute F and then into the ark. Stones were enclosed in the vat or tun. Upon the vat rested the horse which held the hopper & feed shoe. A crook string was fixed to the shoe. The end of the string was positioned near the meal bin on the floor below. It was attached to a peg which could be turned. Pressure on the crook string changed the angle of the shoe & caused more or less grain to fall into the stone's eye. A slide on the hopper H controlled the quantity of grain falling onto the shoe. To keep the shoe pressing against the damsel a wand or spring was fixed to the top of the tun and a string linked them together.

The dressings on the stones were important to the miller. Great danger could arise if the stones ran dry of grain. Without grain the stones would not only lose their dressings but cause a fire. To prevent this happening a bell alarm was contrived. This simple device was a bell on a string with a leather flap at its other end. The latter sat in the hopper covered by the grain. If the hopper ran low the weight of the bell, positioned at x, pulled the leather free. This caused the bell to fall against the moving damsel and set it ringing. A stop knot kept the bell vibrating by the damsel. In some mills the bell string passed through the stone floor & the bell hung above the spur gear.

ADJUSTMENTS

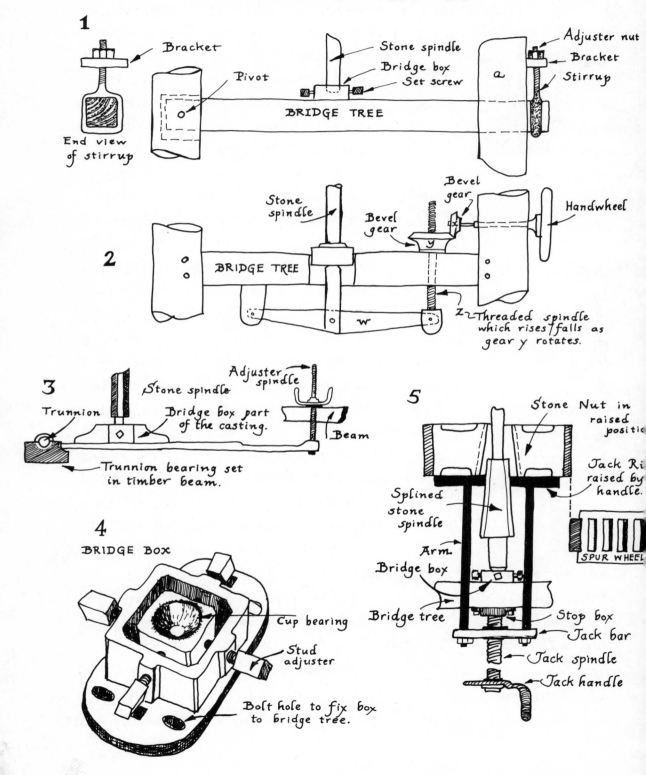

1

Bracket

End view of stirrup

Pivot

Stone spindle
Bridge box
Set screw

BRIDGE TREE

a

Adjuster nut
Bracket
Stirrup

2

Stone spindle

BRIDGE TREE

Bevel gear

Bevel gear

Bevel gear

x

y

w

Handwheel

z. Threaded spindle which rises/falls as gear y rotates.

3

Trunnion

Stone spindle

Bridge box part of the casting.

Adjuster spindle

Beam

Trunnion bearing set in timber beam.

4

BRIDGE BOX

Cup bearing

Stud adjuster

Bolt hole to fix box to bridge tree.

5

Stone Nut in raised position

Splined stone spindle

Jack Ri raised by handle.

SPUR WHEEL

Arm

Bridge box

Bridge tree

Stop box
Jack bar
Jack spindle
Jack handle

During the operation of a mill it was necessary to make adjustments to the stones to regulate the quality of the meal. The distance between the stones could vary as the stone spindle became warm and expanded. A miller had to exercise constant vigilance so that he was aware of the way the mill was grinding. The type of grain being ground and the sharpness of the stones also made a difference to the optimum distance between the stones. Adjustment was made to the stones by raising or lowering the bridge tree ~ a process called TENTERING. As the bridge tree moved the stone spindle moved with it, & the distance between the stones varied. The lower end of the stone spindle rested in a brass cup bearing.

Some methods of tentering are shown opposite. 1. A bridge tree pivoted at one end & resting in a stirrup at the other. From the timber frame a the metal bracket projects to support the adjuster nut. 2. A fixed bridge tree with the cup bearing recessed into it. Adjustment is made by turning the handwheel which alters the position of the bevel wheels ~ x & y. The latter is threaded on the spindle 'z' which raises the beam w thus changing the position of the stone spindle. 3. A metal bridge beam with trunnions ~ like a cannon ~ that allow it to rise or fall. Adjustment is effected by operating the screw control that rests upon a wooden beam. The bridge box 4 with its set screws correctly positioned gave the stone spindle & the runner stone a precise vertical alignment ~ see p.64.

When the mill ceased working each runner stone was disengaged from the spur wheel by operating the jack ring 5. If stones were left in gear and the wheel started accidentally great damage could arise. The stone nut's centre had a coned & splined profile. The nut was raised by turning the jack handle. The threaded jack spindle pushed against the stop box and this caused the arms to move upwards. These arms raised the jack ring which pushed against the underside of the stone nut.

LIFTING THE STONES

One of the most strenuous tasks the miller had to do was to lift the stones for dressing. The average millstone weighed about a ton and had a diameter of about four feet. Before a stone could be lifted the horse and vat had to be removed. Then the damsel could be extracted and the runner was ready to be turned over. The runner stone had two sockets in the skirt at opposite points. A pin or eyebolt could be placed in this cavity and it then provided an anchor for the block & tackle. In later years the pulley blocks used were of steel and they carried chains to bear the load. There were some mills however which retained their wooden blocks & ropes until the 1960s. Turning a stone uppermost required an anchor point for a block or rope above the level of the stone as the sketch opposite shows. Wedges, levers and large sacks of bran to catch the stone falling on its back, augmented the pulley equipment. Many mills had a stone crane which pivotted in a socket in the floor & on a beam. This type of crane had a screw with two hooks that held the caliper arms. The lower end of each caliper had an eye which fitted the trunnions fixed into the stone's skirt. The advantage of this hoist was the ease with which the stone could be turned over.

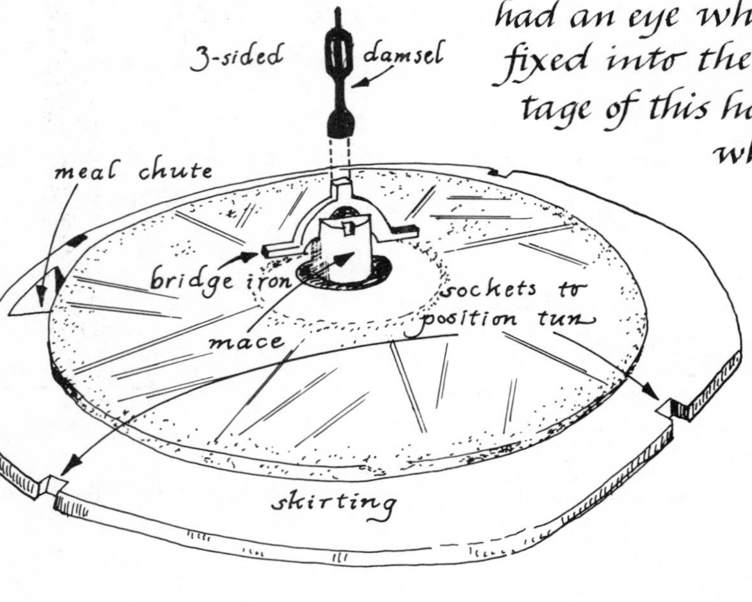

3-sided damsel

meal chute

bridge iron

mace

sockets to position tun

skirting

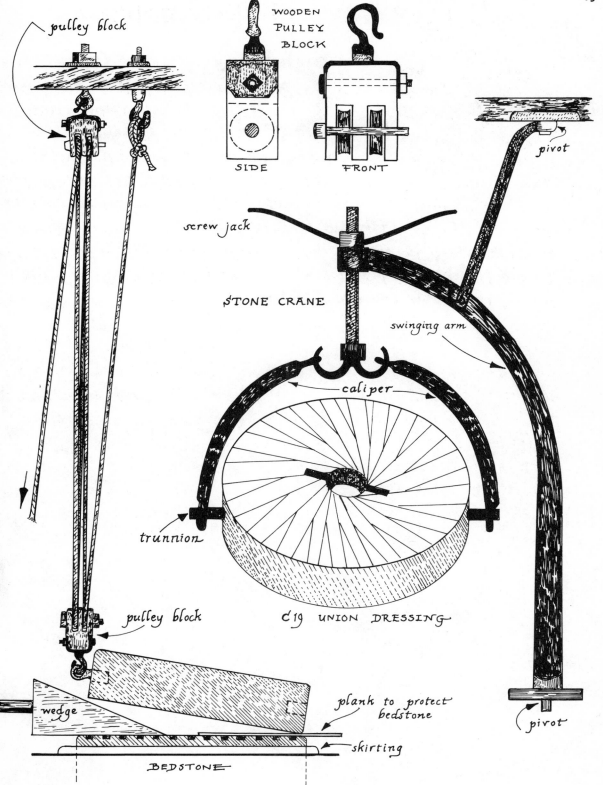

pulley block

WOODEN
PULLEY
BLOCK

SIDE   FRONT

pivot

screw jack

STONE CRANE

swinging arm

caliper

trunnion

pulley block

C19 UNION DRESSING

wedge

plank to protect
bedstone

skirting

BEDSTONE

pivot

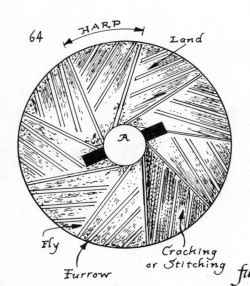

HARP

Land

Fly

Furrow

Cracking or Stitching

A

STONE DRESSING

he unseen faces of the millstones were central to the mill's purpose. It was at this point that the finest adjustments were made. Each stone had its pattern of grooves. These ridges & furrows were very important. The flat surface between the furrows, the lands, contained lesser furrows, stitches, which performed the grinding action. The flour, once ground, fell into the furrows where it worked its way down the slope of the furrow towards the stone's outer edge.

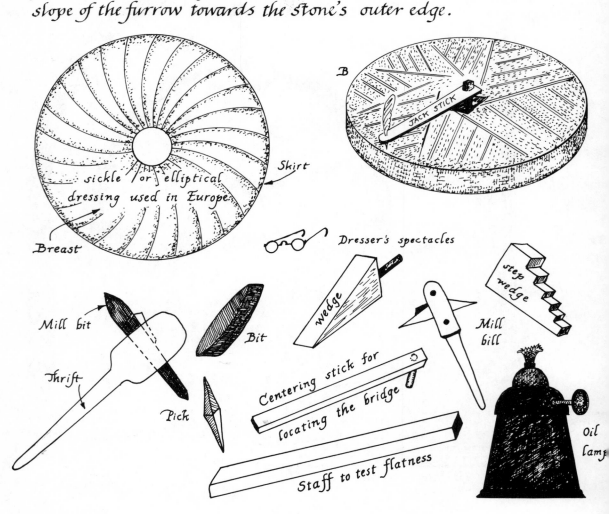

sickle or elliptical dressing used in Europe.

Skirt

Breast

B

JACK STICK

Dresser's spectacles

Mill bit

Thrift

Pick

Bit

wedge

Centering stick for locating the bridge

Staff to test flatness

step wedge

Mill bill

Oil lamp

Dressing a millstone was a skilled task although the tools used were simple. A sharp chisel ~ a bill~ was held in a thrift by a wedge. With this device the stone dresser formed the furrows & stitches which determined the quality of the flour. There were several styles of dressing. The one most often found in England was the common dressing A. Each stone was divided into several harps, & each harp usually had three furrows. It was important for the stone to be flat & this was achieved by using a staff smeared with red oxide. When the staff was rubbed upon the stone's surface the high spots were marked by the oxide. These could be levelled with the bill. A stone dresser protected his eyes by wearing spectacles.

Stones were not always uniformly hard, & in use would often wear unevenly. Apart from needing a flat surface they had to run true on a stone spindle without rocking. If this happened whole grain would emerge from one side & overground meal on the other. The even running of the spindle was checked by using a jack-stick which fitted over the spindle's upper end B. At the outer end of the tracer bar an upright quill was fixed with its lower point touching the stone's surface. As the spindle was turned from below the jack~stick swept across the stone & the quill defined its even or un-even course. To achieve total accuracy adjustments were made to the bridging box ~ p. 60. Two kinds of stone were used. French Burr stones were used for flour. These were made from irregular blocks cemented together & bound with iron straps. Barley was ground with single piece Derbyshire Peak stones.

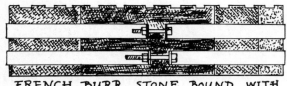

FRENCH BURR STONE BOUND WITH IRON STRAPS

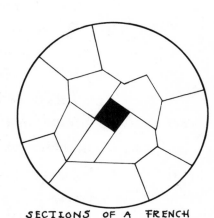

SECTIONS OF A FRENCH BURR STONE

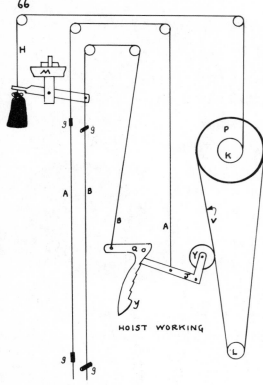

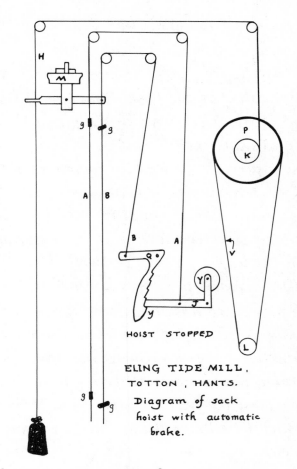

HOIST WORKING

HOIST STOPPED

ELING TIDE MILL,
TOTTON, HANTS.
Diagram of sack
hoist with automatic
brake.

HOISTS

Sack hoists were driven from the crownwheel ~ p.57 ~ which turned a pulley on a layshaft. The driving belt was left slack until the hoist worked. Then the belt had to be brought into tension. Hoists all had characteristics made to fit the layout of a specific mill. Millwrights displayed much ingenuity in their making. To the miller it was an important time and labour-saving device.

The jockey wheel system shown above has an automatic brake. To bring the wheel Y into contact with the belt v the miller pulled rope A. This raised lever J and pushed the jockey wheel against the belt. As the lever moved upwards (from y), its end was held in place by the locking quadrant Q. In this position the hoist chain H was wound onto the barrel K. As the sack reached M it pushed

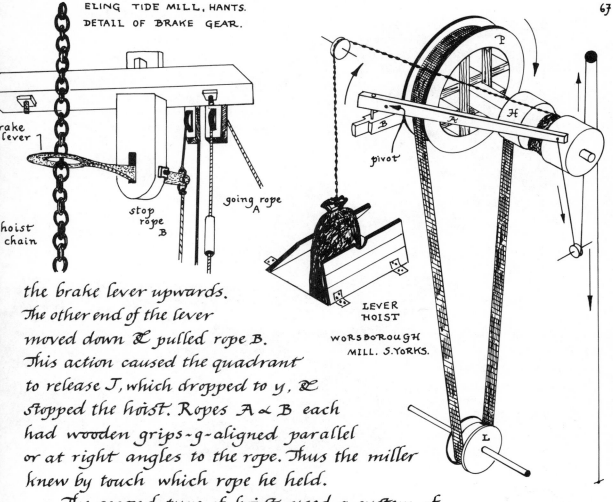

ELING TIDE MILL, HANTS.
DETAIL OF BRAKE GEAR.

brake
lever

hoist
chain

stop
rope
B

going rope
A

pivot

LEVER
HOIST

WORSBOROUGH
MILL. S.YORKS.

the brake lever upwards.
The other end of the lever
moved down & pulled rope B.
This action caused the quadrant
to release J, which dropped to y, &
stopped the hoist. Ropes A & B each
had wooden grips - g - aligned parallel
or at right angles to the rope. Thus the miller
knew by touch which rope he held.

The second type of hoist used a system of
pulleys and levers to tension the driving belt. In
the drawing above a rope running over two
pulleys translates a downward pull into an upward
motion; as the lever A moves it causes the beam B to
rise and tension the belt. The pulley P, which is on the
same spindle as the hoist H, is then turned by the
driving pulley L that is powered by the crownwheel.

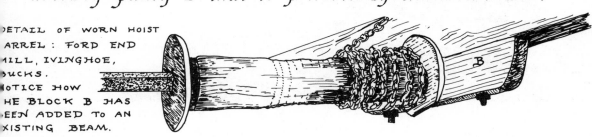

DETAIL OF WORN HOIST
BARREL: FORD END
MILL, IVINGHOE,
BUCKS.
NOTICE HOW
THE BLOCK B HAS
BEEN ADDED TO AN
EXISTING BEAM.

TIDE MILLS

WOODBRIDGE, SUFFOLK. PONDSIDE VIEW
SHOWING THE PROJECTING LUCAM WHICH
CONTAINED THE SACK HOIST. THIS TERM
DERIVES FROM THE OLD FRENCH
'LUCARNE' MEANING A DORMER
WINDOW. THE DOUBLE PITCH
MANSARD ROOF TAKES
ITS NAME FROM
FRANCOIS MANSARD THE
C17 FRENCH ARCHITECT.

ELING MILL, TOTTON, HANTS.
VIEW OF MILL FROM TIDEWAY.
ITS REMAINING WHEEL LIES
INSIDE THE ARCH. THE DOOR
ABOVE ALLOWED BOATS TO
DISCHARGE. THIS IS THE
ONLY MILL NOW REGULARLY
WORKED.

ST. OSYTH, ESSEX. VIEW OF MILL
c.1910. THE EXTENSION, LEFT,
ALLOWED BARGES TO UNLOAD.
THIS BUILDING NO LONGER
EXISTS.

Man discovered that he could harness the immense power of the sea as long ago as the Norman Conquest, C11. There were once many tide mills around our British shores. They were to be found as far apart as West Wales and Woodbridge, Suffolk.

To confine the sea each mill needed a substantial dam. At St. Osyth, Essex the mill pond covered about 30 acres. Each tide mill required the following features : 1. Sea hatches, which worked in one direction only & allowed the rising tide to pass and trapped it when it ebbed. The weight of the captured water closed the hatches. 2. A spillway, weir or tumbling bay which allowed excess water to return to the sea and avoid flooding. 3. A wheelrace to drive the waterwheel.

Each tide provided about four hours milling or more if the pond was large. High tide times change daily and tide millers had to work very irregular hours. The working head of water was created when the tide ebbed leaving the pond full. When the falling tide was low enough the miller could start work. Unlike some rivers tidewater never failed.

DIAGRAM OF WATERCOURSE FOR A TIDEMILL

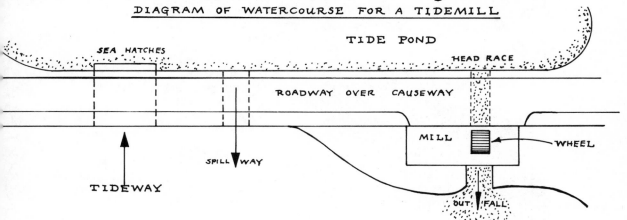

The machinery of a tidemill is the same as other water-mills. The technical distinction arises from the source of the water power.

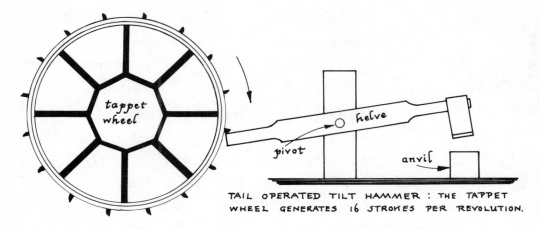

TAIL OPERATED TILT HAMMER : THE TAPPET
WHEEL GENERATES 16 STROKES PER REVOLUTION.

HAMMERING & GRINDING

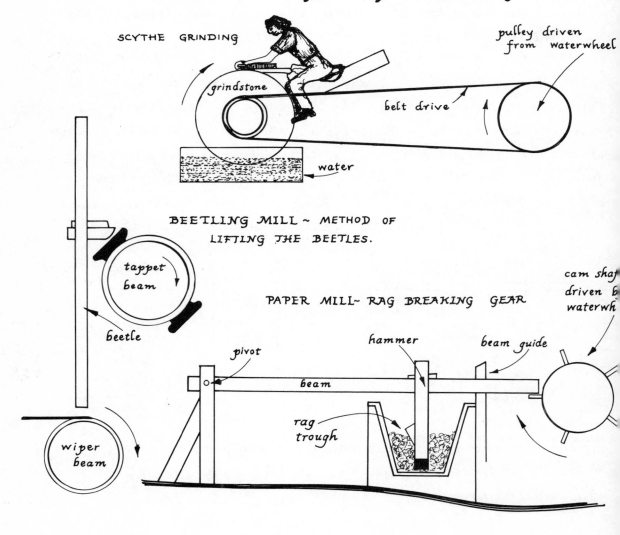

SCYTHE GRINDING

pulley driven
from waterwheel

grindstone

belt drive

water

BEETLING MILL ~ METHOD OF
LIFTING THE BEETLES.

tappet
beam

beetle

PAPER MILL~ RAG BREAKING GEAR

cam shaft
driven by
waterwheel

pivot

hammer

beam guide

beam

wiper
beam

rag
trough

TILT HAMMER – 4 STROKES PER REVOLUTION

The TILT HAMMER appeared at an early stage in the industrial revolution, & its power enabled man to manufacture larger components than ever before. Good examples survive at Sticklepath, Devon; Wortley Top Forge & Abbeydale Hamlet, Sheffield.

The BEETLING MILL was used to finish the surface of linen. As the cloth was wound onto the wiper beam the 32 beetles, lifted by the tappets, fell in turn upon the material.

The FULLING MILL was used to pound the woven cloth into a thicker compressed mass, thus making it stronger & smoother. Fulling mills were used in England from the late C12. These early mills probably had cam operated hammers like the rag mill opposite. By the C19 much heavier fulling stocks were in use.

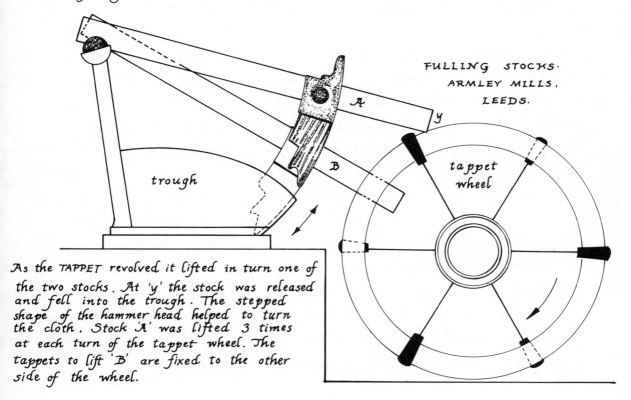

FULLING STOCKS.
ARMLEY MILLS,
LEEDS.

As the TAPPET revolved it lifted in turn one of the two stocks. At 'y' the stock was released and fell into the trough. The stepped shape of the hammer head helped to turn the cloth. Stock 'A' was lifted 3 times at each turn of the tappet wheel. The tappets to lift 'B' are fixed to the other side of the wheel.

GREAT LAXEY MINING COMPANY

THE Lady ISABELLA wheel, at Laxey, Isle of Man, is the most splendid of all the waterwheels in the British Isles. Standing in Laxey Glen it was built to the design of the engineer Robert Casement to pump water from Laxey's deep, 200 fathom mines. This outstanding example of Victorian engineering takes its name from the Governor's wife who was present at the official opening in Sept. 1854. The wheel was used continuously until the mines closed in the depression of 1929. It was purchased by Mr. E.C. Kneale of Laxey who had the foresight to save this engineering treasure, and the largest wheel in these islands, from the hands of the scrap iron merchants. Today thousands of visitors are able to enjoy its splendour and climb the breathtaking stairway to its upper platform.

The wheel is 72ft. 6ins. in diameter and 6ft. wide. It is a pitch back wheel, & the water is drawn from a reservoir via a 2ft. dia. feed pipe. This pipe P is contained inside the stone tower. When water reaches the top it flows beneath the platform and is released onto the wheel at x. The wheel rotates in the direction shown by the arrow Y. To prevent water falling from the buckets the arc s-s is shrouded. Each of the 168 buckets can hold about 29 galls. The wheel's 17ft. long axle is 21ins. in diameter, & weighs about ten tons. At two and a half turns a minute it developed around 200 h.p. which was transferred to the pump by the 2½ ton crank with a throw of 10 feet. A long line of rods linked the wheel to the pump about 200 yards away. To assist the smoothness of the motion a balance box B, once containing some 50 tons of scrap metal, was attached to the southern end of the wheel via a bell crank C worked from the crankshaft D. The wheel could lift 250 gallons of water a minute from a depth of some 200 fathoms.

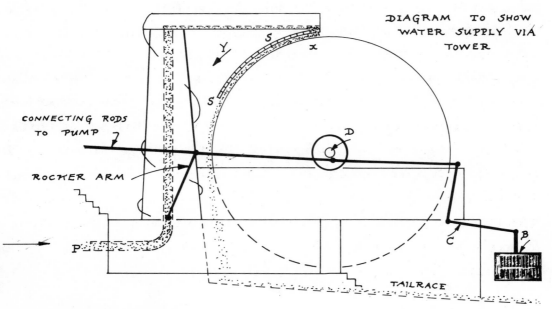

DIAGRAM TO SHOW
WATER SUPPLY VIA
TOWER

CONNECTING RODS
TO PUMP

ROCKER ARM

S

S

Y

x

D

P

C

B

TAILRACE

WATER-POWERED PUMPS

COULTERSHAW PUMP. In 1782 the Third Lord Egremont built a water-powered pump at Coultershaw near Petworth to provide river water to the town. It delivered about 20,000 galls. of water per day. There are three beams which are pivoted at one end ~a. Two rods are attached to the opposite end of each beam. One ~b~ connects the beam to the cranked axle which is driven by the waterwheel. The cranked axle has three journals ~x~ opposed at 120°. An outer plunger rod ~c~ at the end of each beam moves the plunger ~p~ in one of the three cylinders-d. The plungers are back-to-back leather cups. There are leather and brass non-return valves~ y&z. Each cylinder draws its water from the culvert ~e~ which supplies the wheel & expels it via the feed pipe ~f. The present wheel, made at the Cocking Iron Works, mid C19, replaced a wooden one. Each plunger rod is made in two parts joined by a wedged sleeve ~s. This made the task of pump repair easier.

RESTORATION~ Sussex Industrial Archaeological Society. p.154.

MELINGRIFFITH PUMP. Whitchurch, Cardiff. (The prefix 'melin' indicates a mill.) This pump is located on the site of a former water-powered forge. It was constructed by the Glamorganshire Canal Company to lift tailrace water from the adjacent tinplate works into the canal. Two beams each work a pump cylinder. There is an interesting connection between this pump ~1807~ and the Claverton Pump ~1813~ as both were designed by John Rennie. His pump was at work for 135 years until 1942.

RESTORATION~ Oxford House (Risca) Industrial Archaeology Society & The Inland Waterways Association with the City of Cardiff. p.154.

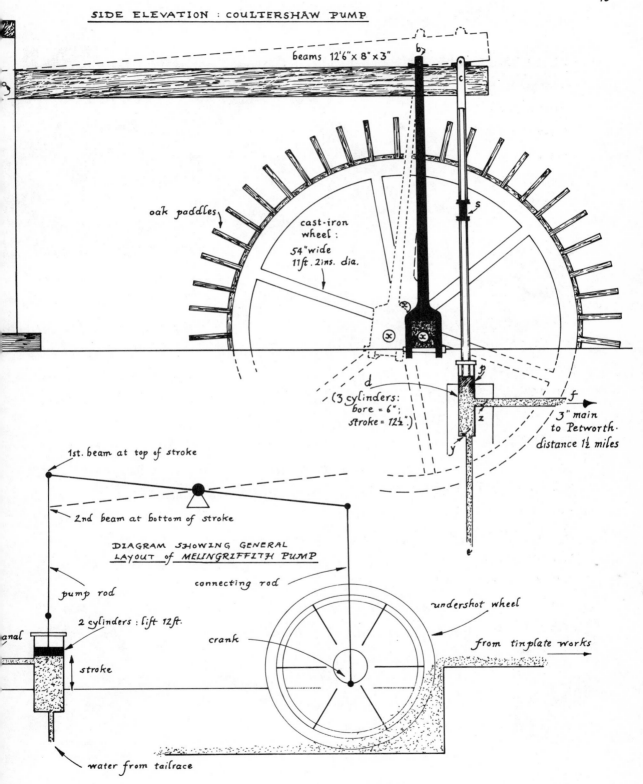

SIDE ELEVATION : COULTERSHAW PUMP

beams 12'6" x 8" x 3"

oak paddles

cast-iron wheel:
54" wide
11ft. 2ins. dia.

d
(3 cylinders:
bore = 6";
stroke = 12½".)

f
3" main
to Petworth.
distance 1½ miles

1st. beam at top of stroke

2nd beam at bottom of stroke

DIAGRAM SHOWING GENERAL
LAYOUT of MELINGRIFFITH PUMP

connecting rod

pump rod

undershot wheel

canal

2 cylinders : lift 12ft.

crank

from tinplate works

stroke

water from tailrace

CLAVERTON PUMPHOUSE ~1813~

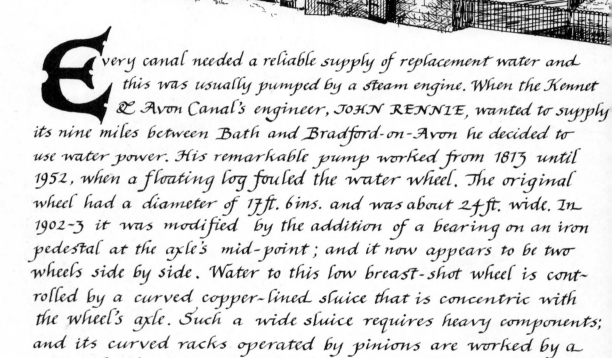

Every canal needed a reliable supply of replacement water and this was usually pumped by a steam engine. When the Kennet & Avon Canal's engineer, JOHN RENNIE, wanted to supply its nine miles between Bath and Bradford-on-Avon he decided to use water power. His remarkable pump worked from 1813 until 1952, when a floating log fouled the water wheel. The original wheel had a diameter of 17 ft. 6 ins. and was about 24 ft. wide. In 1902~3 it was modified by the addition of a bearing on an iron pedestal at the axle's mid-point; and it now appears to be two wheels side by side. Water to this low breast-shot wheel is controlled by a curved copper-lined sluice that is concentric with the wheel's axle. Such a wide sluice requires heavy components; and its curved racks operated by pinions are worked by a series of reduction gears that have a ratio of 171 to 1. Its rebuilt wheel has five tons of elm planking which amounts to 2100 linear feet. In operation about 3½ to 4 tons of water presses upon each of the wheel's paddles.

BUCKET PUMP

coiled rope to form a seal

ONE WAY VALVE

OUT SIDE

INSIDE

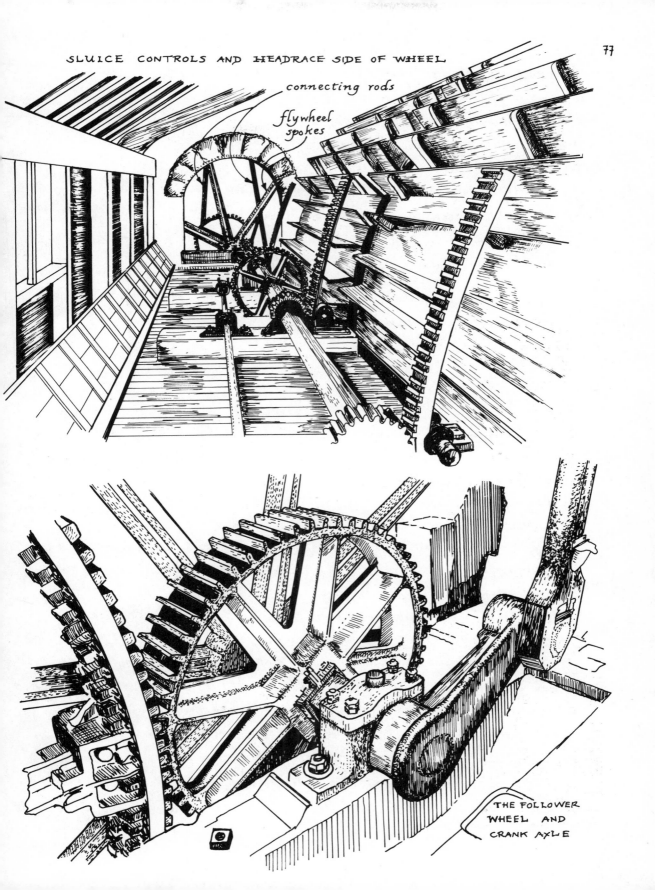

SLUICE CONTROLS AND HEADRACE SIDE OF WHEEL

connecting rods

flywheel spokes

THE FOLLOWER WHEEL AND CRANK AXLE

One of the striking things about the Claverton Pump is the superb quality of its workmanship which has been maintained by those responsible for its skilful restoration. This very remarkable engine works in the following way. The PIT WHEEL is linked to the water wheel's axle by a semi-flexible coupling and turns at the same speed. It is 16ft. dia. & the 12½ inch broad rim has 204 teeth.

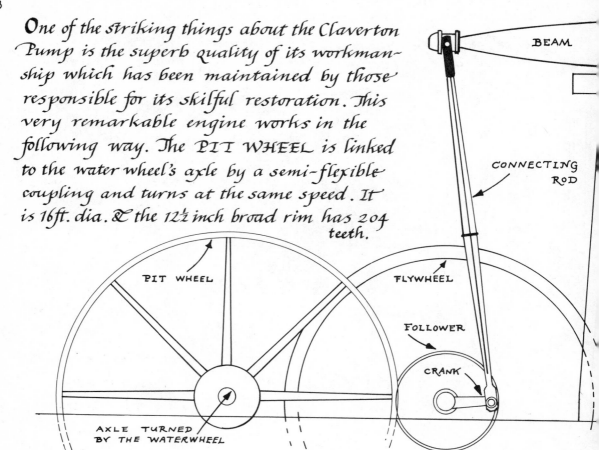

As the PIT WHEEL turns it operates the FOLLOWER, with 64 teeth, which is mounted on the CRANKSHAFT. There are two cranks opposed at 180° with a throw of 4ft. 6ins. The 16ft. dia. FLYWHEEL is also positioned on the crankshaft spindle. It has a rim section 9ins. x 6ins. & its mass helps the evenness of the motion generated. Each crank is linked to one of the two beams by a CONNECTING ROD with a cruciform section & a length of 18ft. Each revolution of the crankshaft rocks the 20ft beams up and down. The oscillations raise & lower the PUMP RODS. This motion sucks in the water through the FILTER, past the one-way valve & into the lower end of the 19 ins. dia. cylinder. The upstroke pulls it via the BUCKET PUMPS into the upper cylinder to be discharged through the OUTFLOW. At the

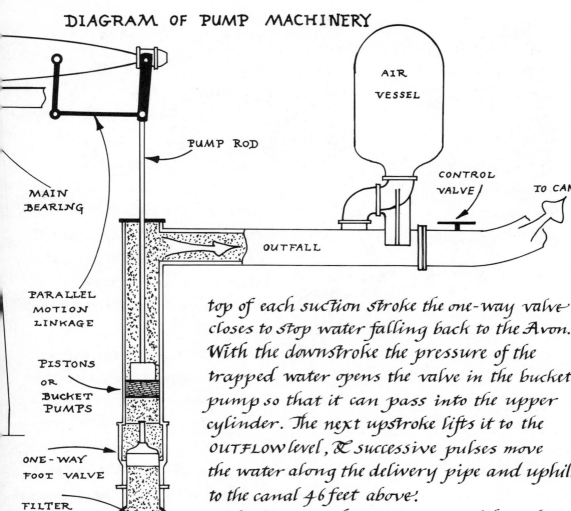

AIR VESSEL

PUMP ROD

CONTROL VALVE

TO CANAL

MAIN BEARING

OUTFALL

PARALLEL MOTION LINKAGE

PISTONS OR BUCKET PUMPS

ONE-WAY FOOT VALVE

FILTER

WATER FROM AVON

top of each suction stroke the one-way valve closes to stop water falling back to the Avon. With the downstroke the pressure of the trapped water opens the valve in the bucket pump so that it can pass into the upper cylinder. The next upstroke lifts it to the OUTFLOW level, & successive pulses move the water along the delivery pipe and uphill to the canal 46 feet above.

On its way the water passes below the AIR VESSEL which absorbs the over-pressure created when the pistons are moving at their top speed ~ 3½ ft. per second. Water then enters the air vessel to be returned to the delivery pipe as piston speed diminishes. The effect is to balance the flow of water within the system. At a waterwheel speed of five turns a minute the pumps make 16 strokes, & at this rate some 98,000 gallons per hour can be lifted to the canal. The fact that after more than 170 years this engine is still at work confirms the excellence of its design.

Visitors can still watch the Claverton Pump at work ~ p. 154.

GUNTON PARK SAWMILL

The business of converting timber into planks was for centuries painfully accomplished by those men, called sawyers, who wrestled with tree trunks over saw-pits. The worst task was that of the bottom sawyer who performed the downward cutting stroke and also suffered the falling sawdust.

The casual visitor who watches the slow pace of the turning mill wheel will usually be unaware of the urgent throb of gears within. Immense power can be derived from water but in a corn mill it is not so easy to see the drama of its work for man. In a saw mill there are both the rhythms of the wheel and the saws. At each stroke these slowly devour the great stock of the tree as it moves forward upon its heavy carriage. The great advantage of a mechanical saw was its capacity to hold several saw blades at once. In one operation it could reduce a tree into the desired number of planks. Gunton Park saw-mill is one of England's most fascinating survivors.

Above: Pit sawing.

Right: An amazing sawing device from the C19 U.S.A.

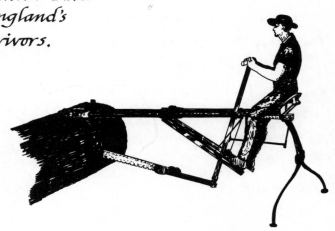

Set in a splendid
landscape, water
from the lake was used
to provide power for this
estate mill: circa 1800.
The machinery is being
restored to a working
state. Visitors will soon

SAW BLADE IN
SHARPENING VICE

be able to marvel at the manner in which Norfolk men set
water to work. The mechanics of this fine mill share elements
also observed in the Dutch mill described on p. 134. The men
who built or maintained these old machines are usually
anonymous, but we do know that the estate carpenter at
Gunton, in 1892, was John Joseph Vince.

The drive from the
waterwheel spur gear · A ·
to the follower · F ·
that operated the
primary pulley · P ·
See next page.

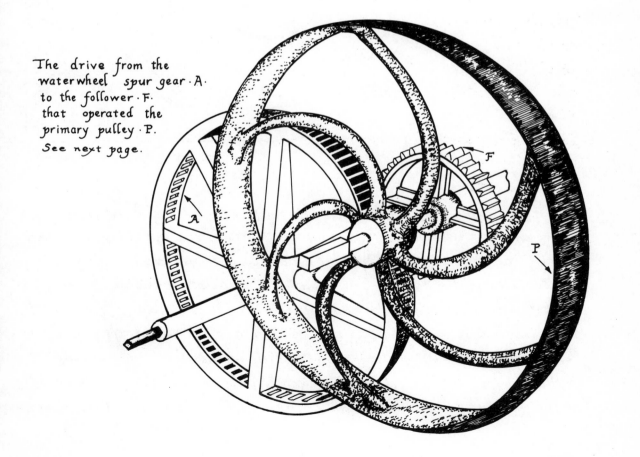

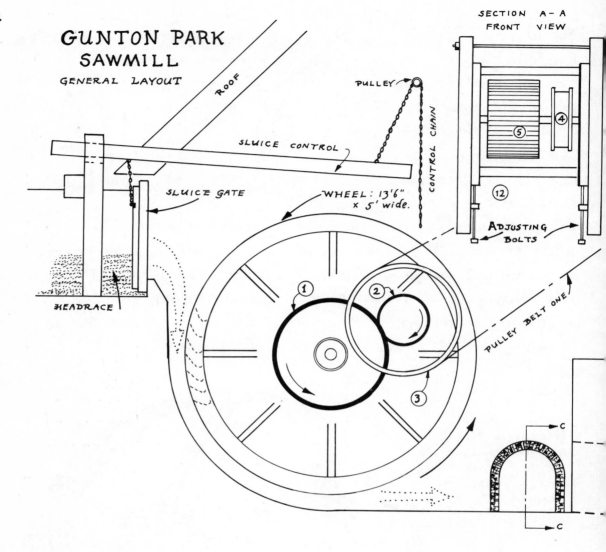

GUNTON PARK
SAWMILL
GENERAL LAYOUT

ROOF

SLUICE CONTROL

PULLEY

CONTROL CHAIN

SLUICE GATE

WHEEL: 13'6"
x 5' wide.

HEADRACE

① ② ③

PULLEY BELT ONE

SECTION A-A
FRONT VIEW

⑤ ④

⑫

ADJUSTING
BOLTS

C
C

One of the impressive aspects of this mill is the size of its components. There are two waterwheels side by side. The second, not shown, worked other machinery. Water from the headrace entered via one of the two sluice gates and struck the wheel at the 10 o'clock position. Expended water was discharged through the tunnel below the mill floor. As the wheel moved so did the spur gear ①. In turn this rotated the follower ② which worked the pulley ③. A belt linked this with the pulley ④ & so the drum wheel ⑤ revolved. The second belt linked the drum wheel to the pulley ⑥ which moved the crank-axle ⑦. Its momentum was

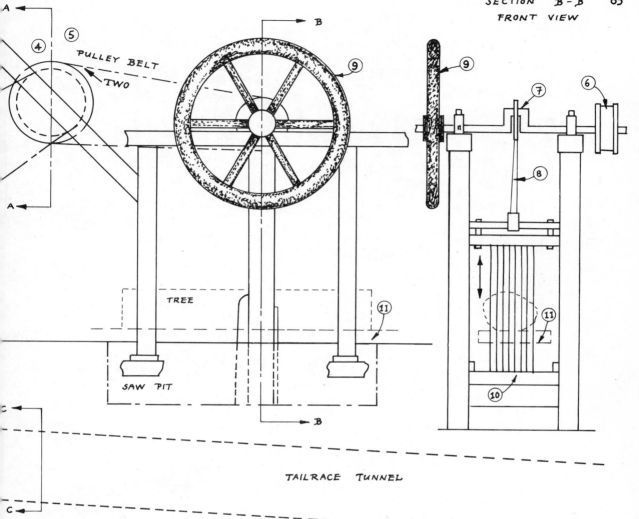

A

5

4

PULLEY BELT

TWO

9

9

7

6

8

B

A

TREE

11

11

SAW PIT

10

B

C

TAILRACE TUNNEL

C

assisted by the robust flywheel⑨ ~ 7ft. 6ins. dia. The crank moved the connecting rod⑧ that joins the saw frame⑩. This moved up & down as the arrows show. This motion moved the timber carriage⑪ against the saw blades. In use belts can stretch & the frame⑫ could be adjusted by bolts.

Spur wheel① is 35 ins. dia & has 96 teeth. The follower② is 24 ins. dia. & carries 36 teeth. Each revolution of the waterwheel turned pulley③ 2·6 times. Saw frame blades are about 85 ins long x 7½ ins. wide and carry about 90 teeth ~ at ¾ inch pitch.

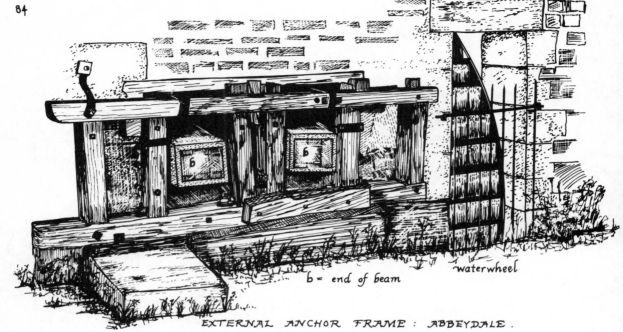

b = end of beam   waterwheel

EXTERNAL ANCHOR FRAME : ABBEYDALE.

A BLOWING ENGINE

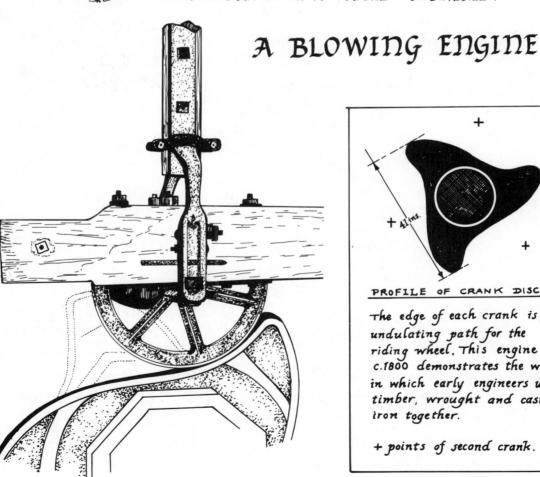

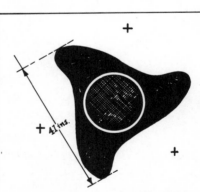

PROFILE OF CRANK DISC.

The edge of each crank is an undulating path for the riding wheel. This engine c.1800 demonstrates the way in which early engineers used timber, wrought and cast iron together.

+ points of second crank.

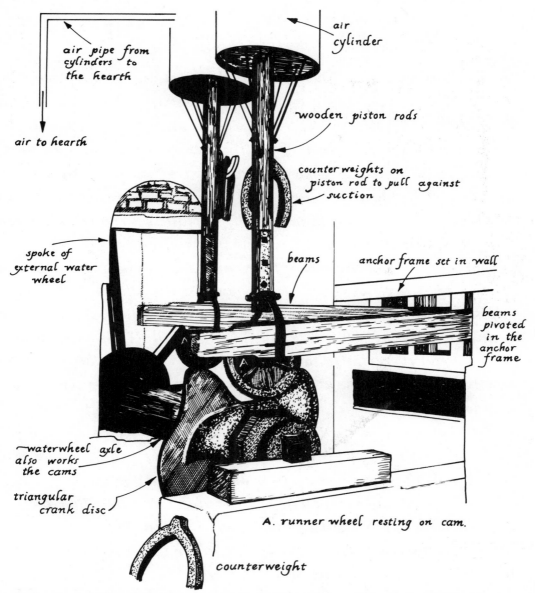

air cylinder

air pipe from cylinders to the hearth

air to hearth

wooden piston rods

counter weights on piston rod to pull against suction

spoke of external water wheel

beams

anchor frame set in wall

beams pivoted in the anchor frame

waterwheel axle also works the cams

triangular crank disc

A. runner wheel resting on cam.

counterweight

GENERAL ARRANGEMENT OF BLOWING ENGINE MACHINERY

AIR was necessary to produce the correct heat at the forge. Country blacksmiths used hand bellows ~ p.26. TILT HAMMER production was a continuous process and a constant supply of blown air essential. At Abbeydale we can still see the massive blowing engine which is operated by its own overshot wheel. The wheel's axle turned the triangular cams. Their shape caused the riding wheels 'A' to move up and down. This enabled the beams to push up or pull down the piston rods. The pistons inside the cylinders sucked in air & then, via the air pipe, fed it to the hearths. Two cylinders provided a constant flow of air.

THE NEEDLE MILL

Needles were some of the first tools devised by man, or perhaps it was an ingenious woman, to enable separate pieces of material to be stitched together. The oldest needles found by archaeologists were made from the bones of animals man had hunted for his food. In time the needle found its place in the sailmaker's loft as well as upon the surgeon's table. There were several stages in needlemaking. The story began in the cottage and the process eventually became industrial.

The final, polishing stage, transforms a needle from a piece of wire into a tool of great precision. Eventually the business of polishing was carried out in mills as a specialist process. At Redditch, Warwickshire, England's only surviving needle mill has been carefully restored and visitors can see the way in which water power was used in the SCOURING or polishing process.

It took considerable skill to polish needles to perfection and the craftsmen who did this work also had to be strong enough to lift the heavy wooden PLATENS on the scouring beds. To give needles their polish they were packed in long canvas or hurden sausages (s), with an abrasive emery powder and soft soap. Hurden was a coarse fabric made from flax or hemp. The sausages were tied tightly with cord and then placed on the scouring bed ~b~ below the platen. Two sausages were laid under each platen and rolled to and fro for five hours. This process was repeated five times. The last time was for glazing the surface when the sausage was packed with vegetable oil and putty powder instead of emery. In this diagram the method of changing the rotary motion of the water wheel to the push-pull action of the platen is shown.

The 14ft. dia. overshot wheel turns a spur wheel A ~ 140 teeth ~ that works the follower B ~ 60 teeth. This latter wheel turns the four throw crankshaft C. A connecting rod D links the crankshaft to the rocker arm E which is called a 'whee-whaw' ~ from its see-saw style of action. At the top of each rocker arm there is a metal pin to which the platen arm G is attached. This arm has a slot at its end which dropped on to the pin H, but it could be quickly released when the platen had to be raised to remove the sausages.

At the Redditch Mill there are four sets of scouring beds. Two are on the ground floor ~ one on each side of the crankshaft. The rocker arm F extends to the upper floor and provides motion for the beds there. Each turn of the waterwheel rotates the shaft C 2·3 times. The platens move to and fro at each throw of the crankshaft, and in operation make about 25 strokes each minute. This is 1500 movements per hour, 7500 in five hours or some 37,000 oscillations during the complete process taking 25 hours.

Tied sausage containing needles.

F

pivot

MOVEMENT   OF E

H

D

C   B

D

A

E

pivot

SIDE VIEW OF LEFT-HAND GROUND FLOOR MACHINERY

STONE CRUSHING

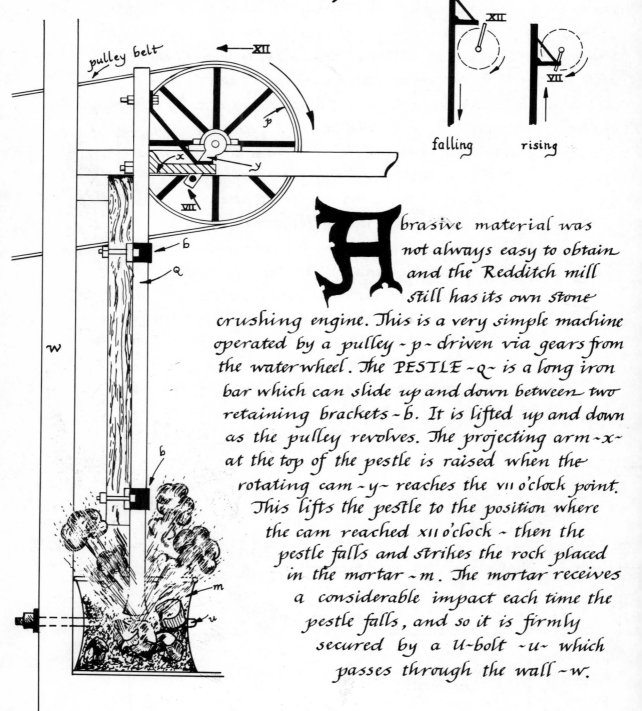

pulley belt

XII

x

y

VII

b

q

w

b

m

u

XII

falling

VII

rising

Abrasive material was not always easy to obtain and the Redditch mill still has its own stone crushing engine. This is a very simple machine operated by a pulley ~p~ driven via gears from the water wheel. The PESTLE ~q~ is a long iron bar which can slide up and down between two retaining brackets ~b. It is lifted up and down as the pulley revolves. The projecting arm ~x~ at the top of the pestle is raised when the rotating cam ~y~ reaches the VII o'clock point. This lifts the pestle to the position where the cam reached XII o'clock ~ then the pestle falls and strikes the rock placed in the mortar ~m. The mortar receives a considerable impact each time the pestle falls, and so it is firmly secured by a U-bolt ~u~ which passes through the wall ~w.

WIND

Long before man discovered how to make the elusive wind work for him he must have wondered at those intricate formations of clouds that wheeled their way across the heavens. It is no surprise to learn that the wind became a symbol for powers unseen but undeniably present ~ Acts 2:2. We cannot say who first had the notion of enlisting the wind to drive a ship, but antiquity has provided us with many examples. Bjorn Landström (The Ship ~ 1961) gives us a valuable clue to the significance of the tree which appears amidships in the Scandinavian Bronze Age

Scandinavia ~ Bronze Age

English C13.

Egypt 3100 B.C.

Scandinavia C8.

Egypt 1200 B.C.

vessel. He tells us of the Finnish practice of carrying leafy branches afloat to use as sails if the wind obliged. Man's skill with sail allowed him to develop the windmill. Mediterranean mills still favour loose unframed triangular sails. The millwrights of western Europe became highly skilled wind-masters.

POST MILLS

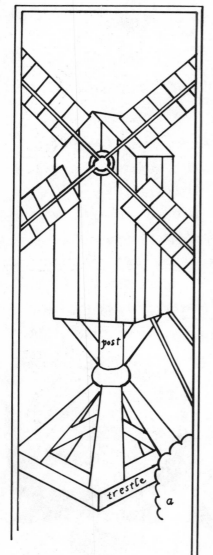

a

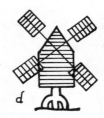

b

c

d

e

f

POST MILLS came into Europe in the C12. probably as a result of ideas borrowed from the Middle East by returning Crusaders. These early mills were simple machines which worked one pair of stones. Our knowledge of these very early mills comes from illustrations made to adorn mediaeval manuscripts & from carvings. Their quality & technical accuracy varies a good deal but there are sufficient common elements to allow us to see that the later mills, apart from being larger, followed the pattern set by the millwrights of old.

The body of a post mill is supported on a vertical post held in place by the members of the trestle. In the drawings the trestle is represented in several ways. All show the sloping timbers which were fixed to the post and to the horizontal beams resting on the ground. The mill bodies have a similar gabled shape. Sails were fixed at the front and the miller went in and out at the opposite end,

where there was a door reached by a ladder. To turn the sails to face the wind the body was moved around the post with the help of a tailpole (c, e, h, j, k). Some mills (k) had a simple yoke ~ a talthur~ fixed to the pole end. The miller placed his head between the staves and pushed. There were four sails with plain lattice frames on which the miller spread his canvas.

As a mill aged it suffered from the wear and tear generated by the weather and the constant vibration of the machinery, during milling. It was common for a mill to begin to lean slightly forward or tilt backwards. The former condition was called headsick, & the latter tailsick. We cannot be sure if the artists drew the mills in this way by accident or as the result of acute observation.

SOURCES: MS. FROM BODLEIAN LIBRARY, OXFORD. a. C15. French. Ms. Douce 276 f 33: b. C16. French. Ms. Douce 135 f 47: c. Carving at Lacey Green mill, Bucks: d. Bench end, Bishops Lydeard church, Somerset: e. C16 English mill on a stone base. Ms. Ashmole 1504 f 14.v: f. C17 carving- exterior Tattenhoe church, Beds: g. Sheldon Tapestry - 1588 - Warwick Museum, Long Compton mill: h. C16 French. Ms. Douce 135 f. 47: j. C15 Bench end, North Cadbury church, Somerset.: k. C14 Flanders. Ms. Bodley 264 f. 81.

TOWER MILLS

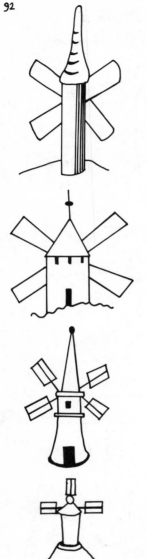

Tower mills, which seem to have appeared in the C14, marked a great technical advance as the miller simply had to turn the cap to face the sails into the wind. Methods of cap turning are detailed below pp. 99~101. Stone, brick and flint have been used to construct tower mills. These materials were more expensive to construct but they had greater durability than timber and required less frequent repairs. In England towers were often made rainproof by adding tar to the outside. This was a functional method, but it did not always enhance a mill's visual aspect.

Knowledge of early tower mills is very sparse and the representations, in glass, manuscripts or carvings, that have survived are therefore particularly valuable. Documentary evidence has to be regarded with some caution, however, as we must always make allowance for the perception of the artist, his skill as a draughtsman and the extent of his technical knowledge. The examples shown vary in their detail. Early illustration is a good area for further research~

MILLS FROM STAINED GLASS
& MANUSCRIPTS ~ C14.

see John Salmon, p. 156. Estate maps which can be found in County Record Offices frequently contain representations of mills, and sometimes record useful technical data.

At Batz, Brittany a mill has survived from the C14. The tower has a very unusual profile. Two balconies give access to the sails. The sailstocks are rather slender, and the sails are without the usual hemlaths. This gives them a fishbone appearance when the sails are undressed.

BATZ, BRITTANY. C14.

ROUMANIA

SMOCK MILLS

In England wooden tower mills are called smock mills as they are said, by those with an active imagination, to resemble the figure of a countryman dressed in a smock. This style of mill originated in Holland in the early C16. Some English examples have white woodwork & others are treated with creosote. One characteristic of the smock mill is its six or eight-sided plan. So many corners were a weakness, and allowed extra points for the rain to come in. To turn the sails into the wind the cap alone had to be moved. This was done with a tail pole - in Holland ; a winch or a fantail - England. The term smock is not used so often in Holland where mills are described by their function - e.g. corn or polder mill. England's most unusual smock mill is at West Blatchington. It stands on top of a range of barns. As well as milling it also worked a thresher & other barn machin.

SANDWICH, KENT.   c.1760

BIDDENDEN, KENT.
NOW DEMOLISHED.

OPPOSITE • DUTCH SMOCK
MILL used for corn grinding ~
Korenmolen. The cap & tower
sides are thatched. Its brick base
provided living space. One pair
of sails is omitted to show how the
cap was shaped to allow for the
backward slope of the sailstocks.

W. BLATCHINGTON,
SUSSEX. c.1820

HORIZONTAL MILLS

In areas where the direction of the wind was more or less constant it was possible to construct mills with sails that worked in a horizontal plane. A tower was built with its 'window' facing the prevailing wind. The principle of catching the wind in this manner was exactly the same as that employed by the millwrights who built the horizontal water wheels - see p.48. Both machines allowed the driving element to impinge on one side of the wheel or sail.

Horizontal windmills in Persia had their stones placed at the base of the tower below an arch that linked two of the walls. A heavy beam was set at the top of the long walls and into its lower side the vertical sail post was fixed. At the bottom of the post (a) the iron mace (b) was placed. The mace was set into a large block of wood (c). This block rested upon a moveable beam (d) which was used for tentering - p.61. It could be adjusted by the wedge (e).

Some horizontal mills were used in England. At the end of the C18 examples were to be seen at Battersea (1788), Margate and Sheerness. Captain Stephen Hooper is credited with the design of the two latter mills - see also p.106.

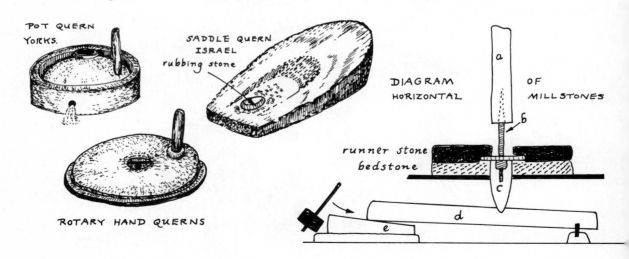

POT QUERN YORKS.

SADDLE QUERN ISRAEL
rubbing stone

DIAGRAM OF
HORIZONTAL MILLSTONES

runner stone
bedstone

ROTARY HAND QUERNS

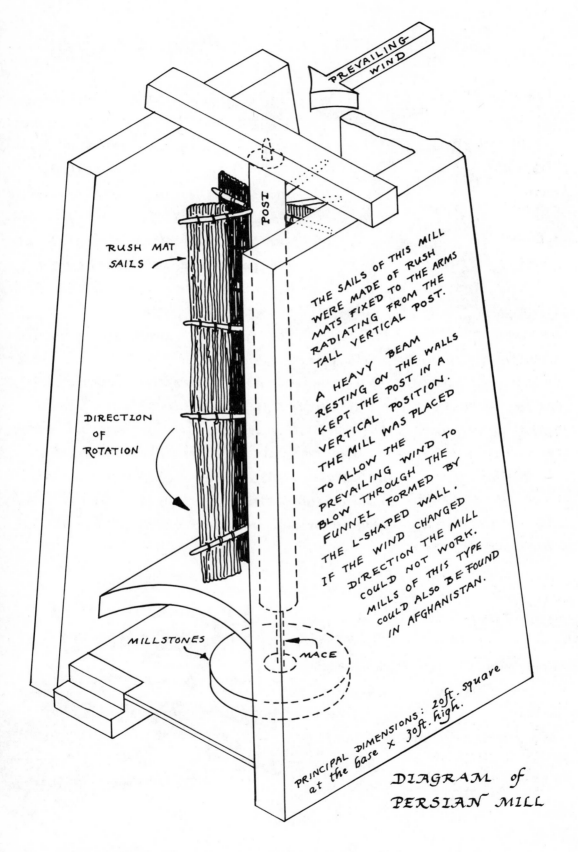

PREVAILING WIND

RUSH MAT SAILS

POST

DIRECTION OF ROTATION

MILLSTONES

MACE

THE SAILS OF THIS MILL WERE MADE OF RUSH MATS FIXED TO THE ARMS RADIATING FROM THE TALL VERTICAL POST.

A HEAVY BEAM RESTING ON THE WALLS KEPT THE POST IN A VERTICAL POSITION. THE MILL WAS PLACED TO ALLOW THE PREVAILING WIND TO BLOW THROUGH THE FUNNEL FORMED BY THE L-SHAPED WALL. IF THE WIND CHANGED DIRECTION THE MILL COULD NOT WORK. MILLS OF THIS TYPE COULD ALSO BE FOUND IN AFGHANISTAN.

PRINCIPAL DIMENSIONS: 20ft. square at the base × 30ft. high.

DIAGRAM of PERSIAN MILL

FAN STAGES

POST MILL with FANCARRIAGE

Fan stages were added to many post mills so that the miller could be spared the demanding task of turning the mill into the wind. Positioned at the rear of the mill the fanwheel was supported on its carriage. When the wind direction changed the fan slowly turned. The gear train linked the fanwheel with the ground wheels that moved the carriage until it reached the point where it was out of the wind and therefore stopped. The sails were then facing in the desired direction. A bevel gear on the fan's axle worked a second bevel wheel. The latter, fixed to a long driving rod, moved a similar gear at ground level. The gears on the ground frame translated the vertical motion of the fan into a horizontal motion which operated the ground wheels. Their width stopped them sinking into the surface.

FAN STAGE or CARRIAGE.
View of the ground wheels showing the heavy timbers used for the framework. One face of each wheel is toothed so that it can be moved by the gear train. A splendid example of a fan carriage can be seen at SAXTEAD GREEN · Suffolk.

See detail opposite ~

FANTAILS

Fan wheels were also used to turn the caps of smock & tower mills. The gear train was linked to a toothed rack on the curb ~ the top of the tower wall

Lincolnshire style cap with fantail.

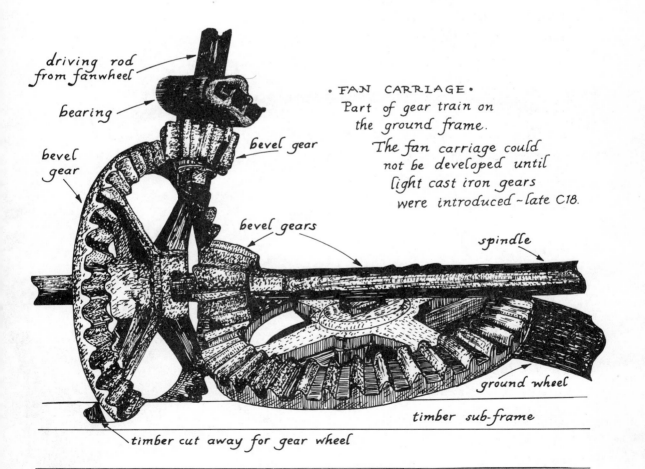

driving rod from fanwheel

bearing

bevel gear

bevel gear

bevel gears

spindle

• FAN CARRIAGE •
Part of gear train on the ground frame.

The fan carriage could not be developed until light cast iron gears were introduced ~ late C18.

ground wheel

timber sub-frame

timber cut away for gear wheel

CAP WINDING

TAILPOLE METHOD
which probably originated in
Holland & was copied in
several other places ~
e.g. East Anglia & South Africa

1. front tie beam : 2. rear tie beam : 3. tailpole :
4. short braces : 5. long braces : 6. capstan :
7. winding rope.

CAPSTAN

The method of winding the cap was the
same as that employed on some post mills.
The winding rope was secured to a post
and the capstan turned by hand. The
tailpole & braces were used as levers to
move the heavy cap. In this drawing you
can see the iron peg 'A' which was
pushed into the tailpole to stop the
capstan being turned.

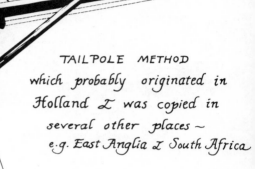

CAPSTAN ~ 'De Kat'
Zaandam.

INTERNAL WINDING WINCH.
Chesterton, Warwicks. [see p. 130].
Probably a C19 addition to
the C17 machinery.

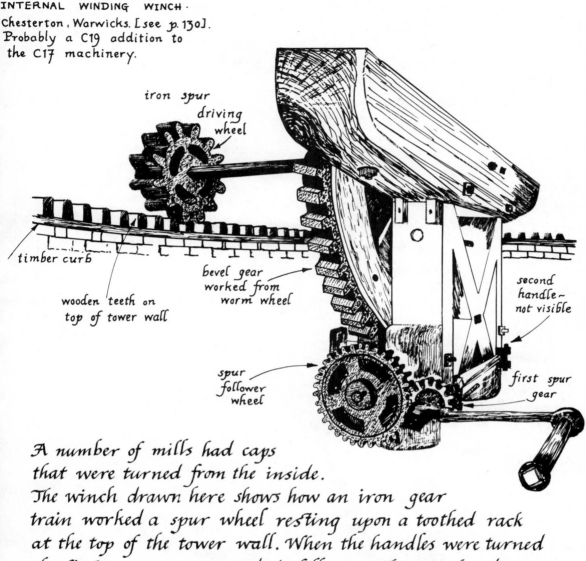

iron spur
driving
wheel

timber curb

wooden teeth on
top of tower wall

bevel gear
worked from
worm wheel

spur
follower
wheel

second
handle ~
not visible

first spur
gear

A number of mills had caps
that were turned from the inside.
The winch drawn here shows how an iron gear
train worked a spur wheel resting upon a toothed rack
at the top of the tower wall. When the handles were turned
the first spur gear moved its follower. This is placed on a
spindle shared by a worm gear ~ not visible. As the latter
moves it slowly rotates the wood and iron bevel gear, which
shares the same spindle as the spur gear positioned on the
wooden rack. To be effective such a device had to be low
geared. Notice how the gear train is arranged with a small
gear working a larger one. Caps could also be turned by a
large exterior pulley & an endless chain ~ e.g. in Anglesey
& on the Fylde.

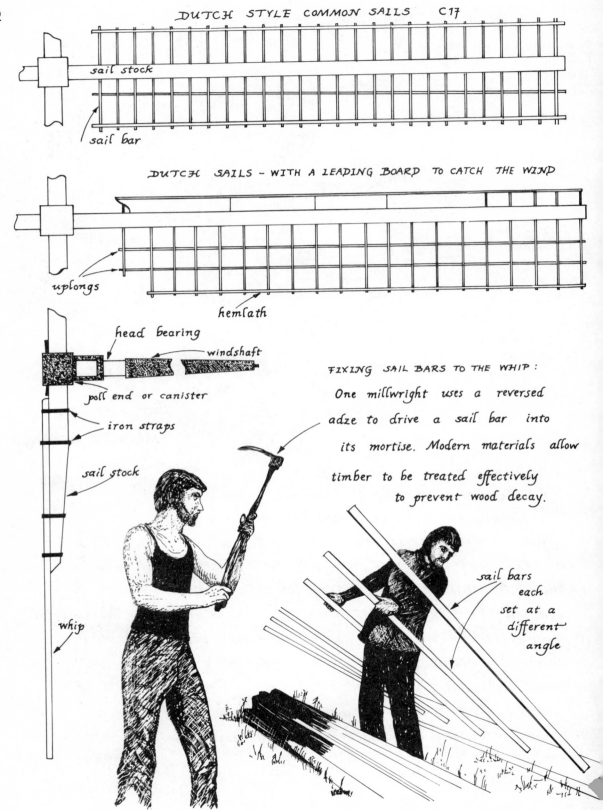

DUTCH STYLE COMMON SAILS   C17

sail stock

sail bar

DUTCH SAILS - WITH A LEADING BOARD TO CATCH THE WIND

uplongs

hemlath

head bearing

windshaft

poll end or canister

iron straps

sail stock

whip

FIXING SAIL BARS TO THE WHIP :

One millwright uses a reversed
adze to drive a sail bar into
its mortise. Modern materials allow
timber to be treated effectively
to prevent wood decay.

sail bars
each
set at a
different
angle

COMMON SAILS

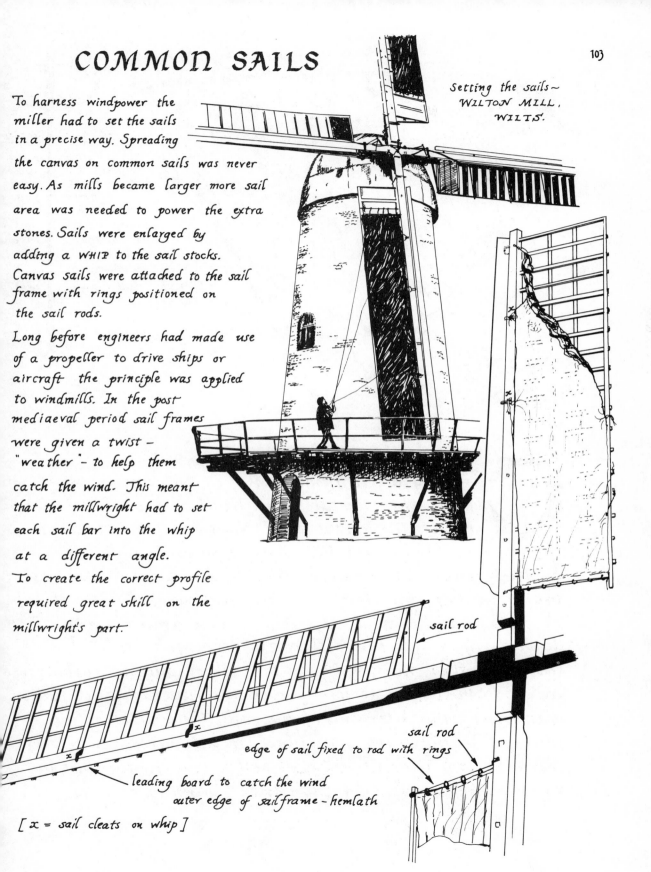

To harness windpower the miller had to set the sails in a precise way. Spreading the canvas on common sails was never easy. As mills became larger more sail area was needed to power the extra stones. Sails were enlarged by adding a WHIP to the sail stocks. Canvas sails were attached to the sail frame with rings positioned on the sail rods.

Long before engineers had made use of a propeller to drive ships or aircraft the principle was applied to windmills. In the post mediaeval period sail frames were given a twist — "weather" — to help them catch the wind. This meant that the millwright had to set each sail bar into the whip at a different angle. To create the correct profile required great skill on the millwright's part.

Setting the sails ~ WILTON MILL, WILTS.

sail rod

sail rod

edge of sail fixed to rod with rings

leading board to catch the wind

outer edge of sailframe – hemlath

[x = sail cleats on whip]

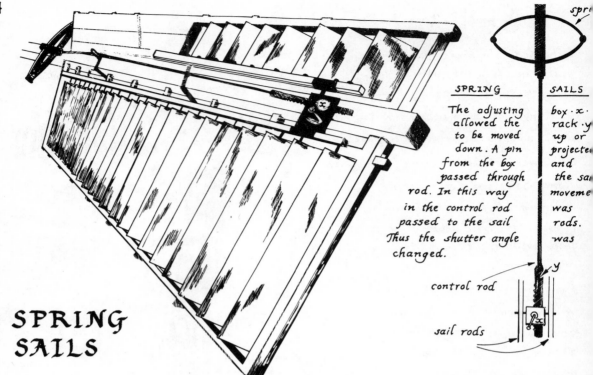

The adjusting
allowed the
to be moved
down. A pin
from the box
passed through
rod. In this way
in the control rod
passed to the sail
Thus the shutter angle
changed.

box·x·
rack·y
up or
projecte
and
the sai
moveme
was
rods.
was

control rod

sail rods

SPRING
SAILS

Climbing a sailframe in wet or icy weather was more
than uncomfortable, and could often be dangerous.
For centuries the miller had been used to doing
this in order to adjust his canvas sails. In 1772 the
Scottish engineer Andrew Meikle devised an alternative
to the cumbersome cloth sail. His sail carried a set of
shutters that could be adjusted to any required angle,
depending on the wind's strength, from ground level. At
last the miller was freed from the task of pulling stubborn
canvas. Shutters were set by operating the rack and pinion
gear -x- at the outer end of the sail. The tension of the
springs held the shutters in place. It was necessary to stop
the mill each time the shutters had to be adjusted, but the
task was much quicker than with flapping canvas.

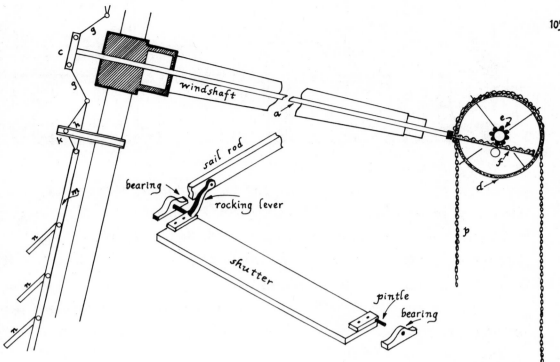

windshaft

sail rod

bearing

rocking lever

shutter

pintle

bearing

PATENT SAILS

In 1807 William Cubbit improved on Andrew Meikle's shuttered sails by contriving a mechanism which allowed the shutters to be adjusted without stopping the mill.

This significant advance was achieved by a series of levers controlled by the striking rod ~a~ which passed through the windshaft & emerged at the centre of the sails. There it was connected to the spider ~c~. The movement of the striking rod was controlled by a chain wheel ~d~ which operated a pinion ~e~ that moved the rack ~f~. The spider moved with the striking rod and its movement adjusted the levers ~g~ and bell cranks ~h~ which were pivotted at ~k~. Thus was motion passed to the sail rods ~m~. Their movement adjusted the angle of the shutters ~n~. Weights were suspended from the striking chain ~p~ to balance the pressure of the wind and keep the blade angle constant. Even bicyles have been used for this purpose!

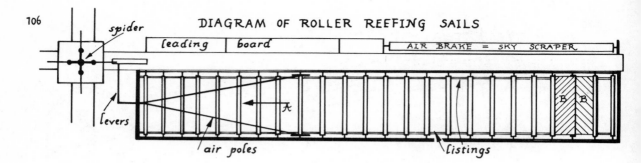

spider

leading   board   AIR BRAKE = SKY SCRAPER

levers

A   B B

air poles   listings

ROLLER REEFING SAILS : In 1789 Capt. Stephen Hooper devised a variation on Andrew Meikle's shutters. He substituted a series of roller blinds which were operated simultaneously by the spider and its levers acting upon the air poles. These opened or closed the blinds, which were linked together by webbing straps called LISTINGS. As the air poles moved in the direction shown by arrow A each blind un-rolled to cover a space ~ B. In use the listings were vulnerable to the effects of the weather, and once broken the blind would not unroll. It was an advantage to be able to adjust the blinds without stopping the mill, but sails of this type do not seem to have been as widely used as Patent sails. The restored mill at Ballycopeland, County Down has sails of this type. A longitudinal shutter on the leading edge acted as an air brake ~ it was also known as a skyscraper.

DEKKERISED SAILS : The Dutch millwright A.J. Dekker (c.1920) invented the streamlining of a sail's leading edge. This device added to a mill's effectiveness. Dekkerised sails did not have whips. The bars were mortised into the sail stocks. Sails of this type were also made of steel.

CAST IRON POLL END · from Mutton Hall Mill, Sussex.
Length = 7'6". Now at W. Blatchington
~ see p. 154.

· iron rings were used to secure fins to the wooden windshaft.

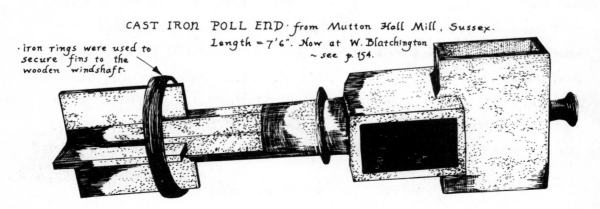

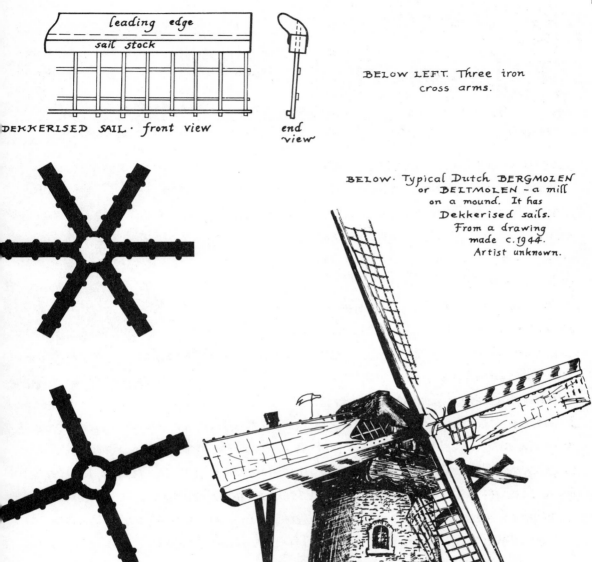

leading edge

sail stock

DEKKERISED SAIL · front view

end view

BELOW LEFT. Three iron cross arms.

IRON CROSS ARMS

BELOW. Typical Dutch BERGMOLEN or BELTMOLEN – a mill on a mound. It has Dekkerised sails. From a drawing made c.1944. Artist unknown.

MULTI-SAILS

There came a time when millers began to understand that the power generated by the sails had something to do with the area presented to the wind. They started to devise ways of increasing the sail area. There was a limit to a sail's width, so they looked for ways of adding to the number of sails. Before the introduction of cast iron it was difficult to place more than four sails on a timber

ENGLAND'S ONLY EIGHT-SAILER. THE TYPICAL TARRED TOWER AT HECKINGTON, LINCS.

windshaft. A mortise made in any timber weakens the structure and heavy sail stocks require considerable support. The use of the cast iron cross arm which could be fixed to an iron windshaft solved many of the technical difficulties involved in timber components. The mill at Heckington, Lincs. is the only surviving eight-sailed mill in England. There were once several six-sailed which included some post as well as tower mills. Five-sailed mills can still be seen at Alford & Burgh-le-Marsh, Lincs. Five-sailed mills were not so easy to work if one sail became damaged. Then the sails would be out of balance.

Multiple sails were not confined to Europe. The Minorcan mill opposite has six sails. In south west Russia, Bessarabia, the post mills had a different shape; although they were still turned by a tail pole.

WORKING FIVE-SAILER
ALFORD, LINCS.

MINORCA
SIX-SAILED
TOWER MILL

RELICT
-SAILER AT
G SUTTON,
NCS. ONCE
GLECTED, MILLS
L SOON DECAY.

POST MILL - BESSARABIA

BRACKES

iron
fixed to

strap
cap frame

DUTCH BRAKE LEVER MECHANISM

were many reasons for stopping a mill. Once the sails were in motion considerable forces were at work, and a brake mechanism had to be very robust to be effective. Millwrigh made brake beams heavy so that their dead weight brought the mill to a halt. Windmill brakes were made from several curved wooden segments, linked together by iron straps, which surrounded the rim of the brakewheel. In these sketche the brake wheels are not shown so that the details of the mechanism can be more clearly seen. When the brake was ON and preventing the mill from turning the brake beam ~b~ was free to fall and tighten the grip of the brake sections on the rim of the brake wheel. The brake was released by pulling the rope ~r~ which caused the brake beam to rise and

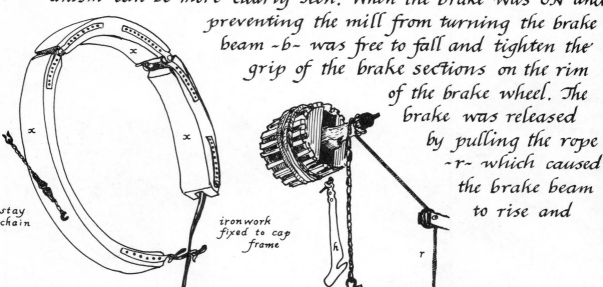

stay
chain

ironwork
fixed to cap
frame

DUTCH PULLEY OPERATED BRAKE. *After Anton Sipman.*

engage the hook ~h. The brake wheel could then move freely inside the rim brake ~x. Some mills had a brake lever ~m~ which extended outside the cap and allowed the brake to be operated from the gallery or ground level.

In place of a lever, pulley systems were also used ~ below & opposite. A brake in the OFF position could be applied by pulling on the brake rope so that the beam moved up & released itself from the hook ~h.

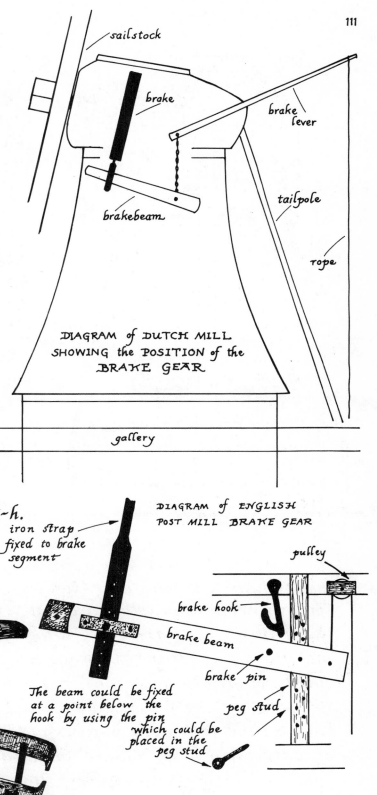

DIAGRAM of DUTCH MILL SHOWING the POSITION of the BRAKE GEAR

sailstock

brake

brake lever

brakebeam

tailpole

rope

gallery

iron strap fixed to brake segment

DIAGRAM of ENGLISH POST MILL BRAKE GEAR

pulley

brake hook

brake beam

brake pin

peg stud

The beam could be fixed at a point below the hook by using the pin which could be placed in the peg stud

Iron strapwork to link brake segments.

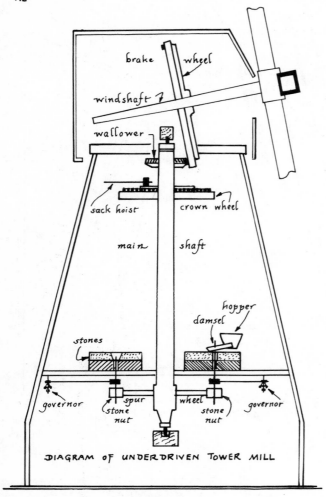

brake   wheel

windshaft

wallower

sack hoist   crown wheel

main   shaft

hopper

damsel

stones

governor   spur   wheel   governor
stone   stone
nut   nut

DIAGRAM OF UNDERDRIVEN TOWER MILL

UNDERDRIVEN STONES

When a mill has underdriven stones the driving power from the windshaft is transferred to the mainshaft via the brake and wallower wheels. The spurwheel then works the stone nuts which provide the final drive to the stones. An underdrift stone's working details are the same as those of a typical watermill ~ p.58; but a jack ring ~ p.57 ~ is not used to put the stones out of gear. In an underdrift windmill the miller has to depend upon the brake.

TENTERING

Of all the machinery in a mill the finest adjustments had to be made between the grinding faces of the stones. The runner and bedstone did not touch, but the distance between them was carefully regulated by the tentering gear. To maintain a constant distance between the stones, and a regular quality of grinding, a pulley on the stone spindle was linked to the governors by a belt. If the speed of the mill increased the balls on the governors moved outwards ~ as a consequence of centrifugal force. This action made the forked end of the steelyard in the collar to rise. Its opposite end x moved

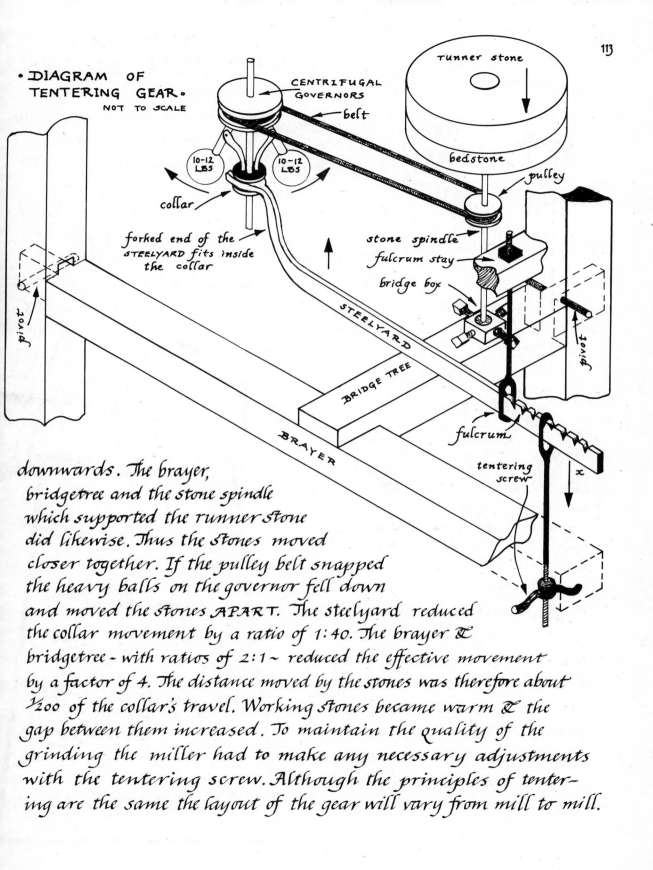

• DIAGRAM OF TENTERING GEAR.
NOT TO SCALE

CENTRIFUGAL GOVERNORS

belt

runner stone

bedstone

pulley

10-12 LBS   10-12 LBS

collar

forked end of the STEELYARD fits inside the collar

stone spindle

fulcrum stay

bridge box

pivot

STEELYARD

BRIDGE TREE

BRAYER

fulcrum

tentering screw

pivot

x

downwards. The brayer,
bridgetree and the stone spindle
which supported the runner stone
did likewise. Thus the stones moved
closer together. If the pulley belt snapped
the heavy balls on the governor fell down
and moved the stones APART. The steelyard reduced
the collar movement by a ratio of 1:40. The brayer &
bridgetree - with ratios of 2:1 - reduced the effective movement
by a factor of 4. The distance moved by the stones was therefore about
1/200 of the collar's travel. Working stones became warm & the
gap between them increased. To maintain the quality of the
grinding the miller had to make any necessary adjustments
with the tentering screw. Although the principles of tentering are the same the layout of the gear will vary from mill to mill.

OVERDRIVEN STONES

The stones of a windmill can be driven from above or below. Early post mills had a single pair of stones driven from above by a lantern pinion and a quant post. This arrangement of gears is called OVERDRIVEN. As mills became larger more

DIAGRAM OF OVERDRIVEN POST MILL
a: windshaft. b: brakewheel. c: lantern pinion. d: quant post. e: stones. f: brake beam. g: brake rope. p: main post. q: quarter bars. r: cross tree. s: piers.

than one pair of stones could be overdriven by a common spurwheel and stone nuts mounted on a quant. The upper end of a quant was housed in a bearing, but it could easily be released to put the stones out of gear. This was an essential

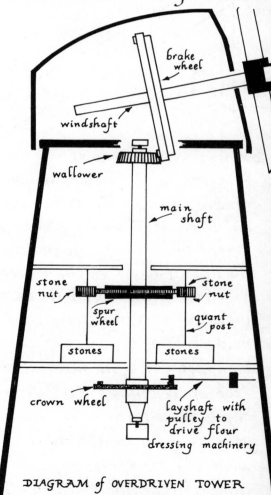

DIAGRAM of OVERDRIVEN TOWER MILL

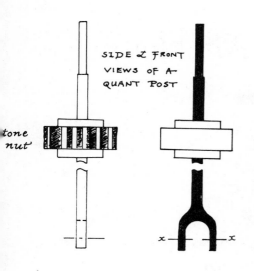

SIDE & FRONT
VIEWS OF A
QUANT POST

tone nut

Runner stone

z

Gymbal

x

Mace

y

edstone

Pulley for governors

Stone spindle

Bridge box

precaution when the mill was not at work. Quant is a word from the C15. It was the name given to the pronged pole used by bargemen in eastern England to propel their boats. This word may derive from the Latin ~ contus = boat pole. When the stones are assembled it is not possible to see the separate parts which allow the runner stone to work without rocking. At work the stones do not touch. The minute space between the moving runner stone and stationary bedstone determines the quality of the flour produced. In the diagram the working parts of overdriven, or overdrift, stones are shown. The bedstone rests upon the heavy structural timbers, & through it passes the stone spindle which must be in an exact vertical alignment ~ see p.64. The upper end of the square stone spindle ~y~ fits into the mace. When the mace is in position the gymbal fits into it and supports the runner stone on its arms ~z. These fit into the grooves cut into the underside of the runner stone as shown. Two slots at the top of the mace ~x~ admit the forked ends of the quant pole. As it rotates with the stone nut, driven by the spurwheel, the quant turns the mace & thus the stone. A pulley on the stone spindle operates the centrifugal governors which determine the space between the stones.

AN ENGLISH WIND PUMP

Many aspects of technology were commonly applied in England and the Netherlands. One of the reasons for finding similarities between English and Dutch practice dates from the C17. At that time that remarkable Dutch engineer, Cornelius Vermuyden, carried out his monumental work of draining the Fenland. To accomplish this, immense new dykes were cut and many wind-driven pumps were built to discharge the surplus water. The landscape was, during some twenty-six years, transformed and the wind's contribution should not be forgotten.

VERMUYDEN's specifications no doubt followed Dutch practice. It is not at all surprising to find that some of the Dutch ideas lasted down the years and became part of an English tradition. This drawing shows the features of the wind engine shown in Walter Blith's "England's Improvement" - 1652. The sails drive a brakewheel. This motion is then transferred to the vertical shaft by a

The Mill made open that the whole Engin may appear

After Walter Blith ~ 1652.

lantern pinion. Lower
down a larger lantern
wheel turns the scoop
wheel - via a peg wheel.

As the scoop wheel
turns it lifts the water
into a higher channel.
The smock mill at Wicken
Fen, the National Trust
nature reserve, shows
several Dutch features.
Its small cap has a typical
Dutch tailpole with the
characteristic beam pro-
truding from each side.
The scoop wheel is driven
by iron gears, but the mill works on the same principles
shown by Walter Blith.

The scoop wheel below has a shallow lift. The long paddles
push water up the slope. When the water reaches the threshold it spills into the outfall. The threshold was raised to
stop water falling back to the lower level. Although each
blade delivered a small
quantity of water pumps
were effective machines.

SCOOP WHEEL DIAGRAM

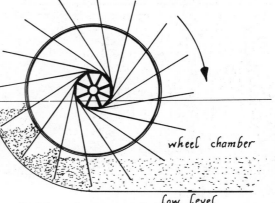

outfall

threshold

wheel chamber

The brick wheelchamber was
very narrow to stop water
spilling back around the
sides of the rising blade.

low level

WIND PUMPS

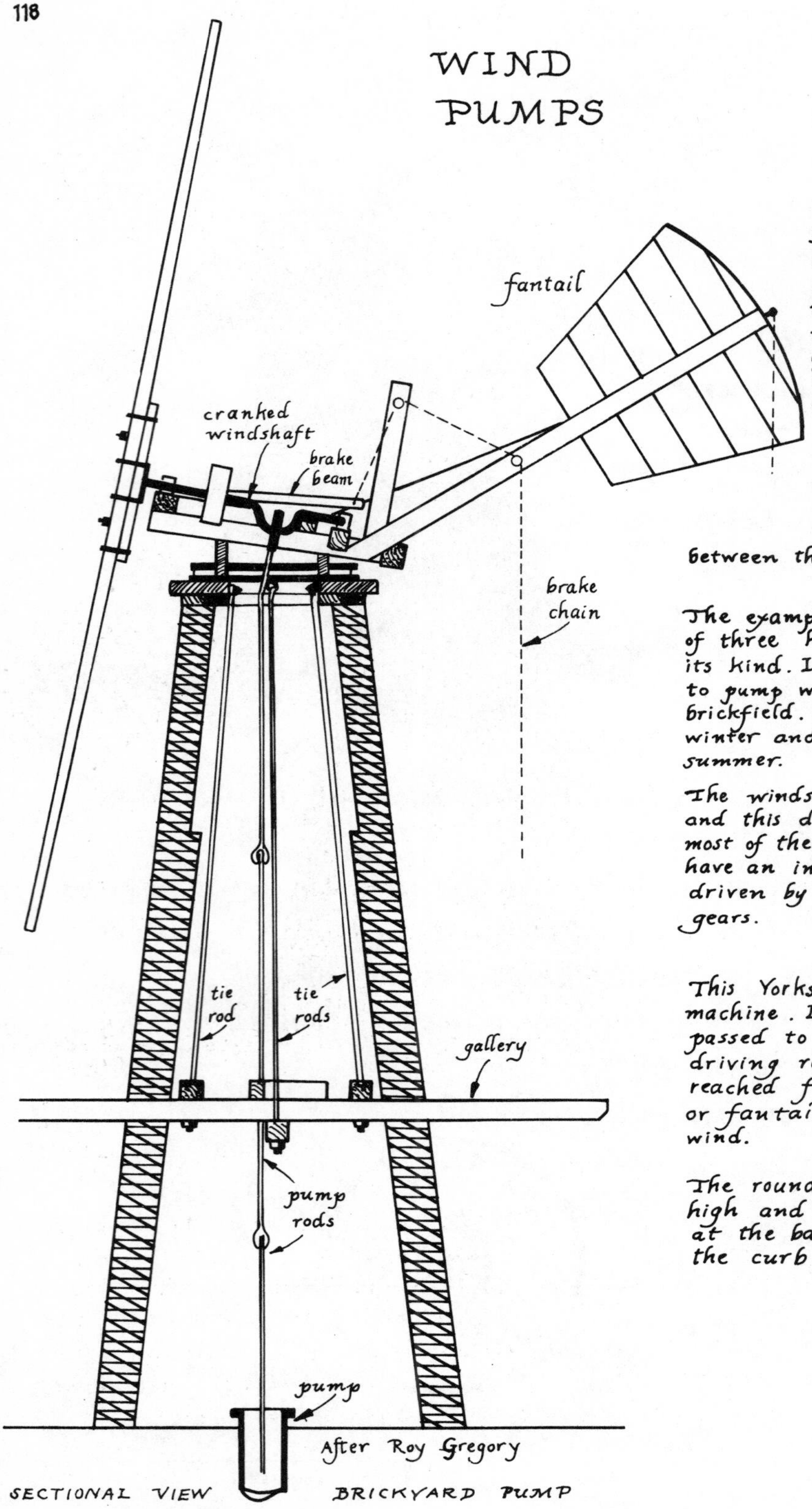

fantail

cranked windshaft

brake beam

brake chain

tie rod

tie rods

gallery

pump rods

pump

After Roy Gregory

SECTIONAL VIEW   BRICKYARD PUMP
HOWDEN : YORKS.

A true pumping action as opposed to a water lifting device on the previous page needs a forward & reverse motion. In windmills the rotation of the windshaft was converted into a push-pull action with a crankshaft.

In the two mills shown here we can see a single and a three cylinder pump. There is an interesting difference between these two mills.

The example shown on this page is one of three known surviving pumps of its kind. It was used in East Yorkshire to pump water from the clay pits in a brickfield. Brickmakers dug clay in the winter and made bricks in frost free summer.

The windshaft of this mill is cranked and this distinguishes the type from most of the other pumping mills which have an intermediate crankshaft, driven by a conventional train of gears.

This Yorkshire pump is a very simple machine. Its oscillating motion is passed to the pump by a series of driving rods. The sails could be reached from the gallery. A vane or fantail kept them facing into the wind.

The round brick tower is about 20ft. high and almost 10ft. in diameter at the base. Four tie rods anchor the curb to the timbers of the gallery.

EASTBRIDGE WIND PUMP : from Minsmere Level, Leiston, Suffolk.
Re-erected at the
MUSEUM of EAST ANGLIAN LIFE,
Stowmarket, Suffolk.

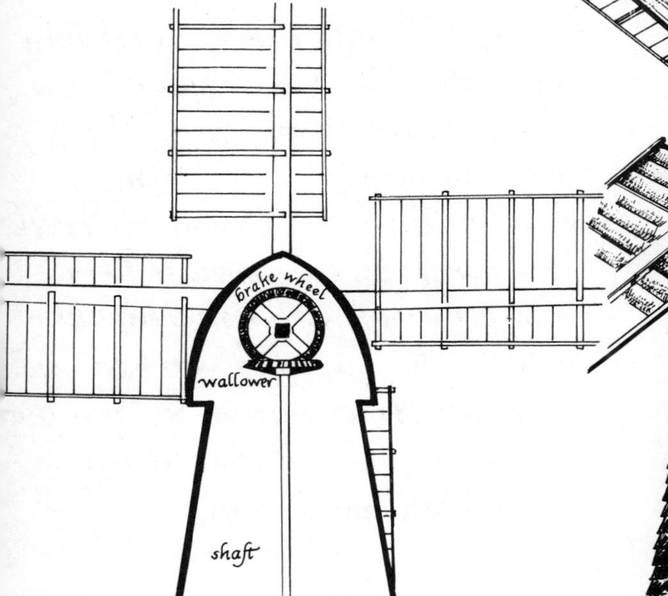

brake wheel

wallower

shaft

bevel gear

crank shaft

connecting rods

OUTFALL

pump

INLET

Sectional view of pump.
The water is lifted 6ft.

Inside the smock tower the brake-wheel drives the wallower & the vertical shaft. This works the three-throw crankshaft. Connecting rods link the crankshaft to the pump. As the cranks are set at 120° to one another the three cylinders provide a constant flow of water. The patent sails are turned into the wind by a fantail - see page 131.

HOLLOW POST MILLS ENGLAND

There is little doubt that the hollow post mill design came from the Netherlands probably via Cornelius Vermuyden.

This example was formerly used to pump water from a brickfield on the Pevensey Levels. Its large circular vane kept the common sails facing the wind. The pump is worked by two eccentric wheels~x. Thompso of Lewes manufactured the cast iron pump~y.

HOLLOW POST MILL from PEVENSEY LEVELS

RE-ERECTED at WEALD & DOWNLAND OPEN AIR MUSEUM, SINGLETON, SUSSEX.

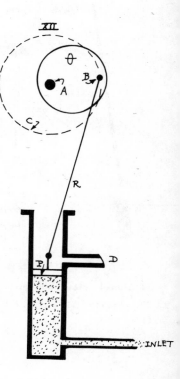

ECCENTRIC OPERATED PUMP: An eccentric provides the same push-pull effect as a cranked axle. As the disc θ revolves around axle A the point B describes the circle C. The rod R is also fixed to the piston P. P moves up and down as B rotates. This action sucks in the water which discharges at D when B reaches XII.

POLDER MILLS

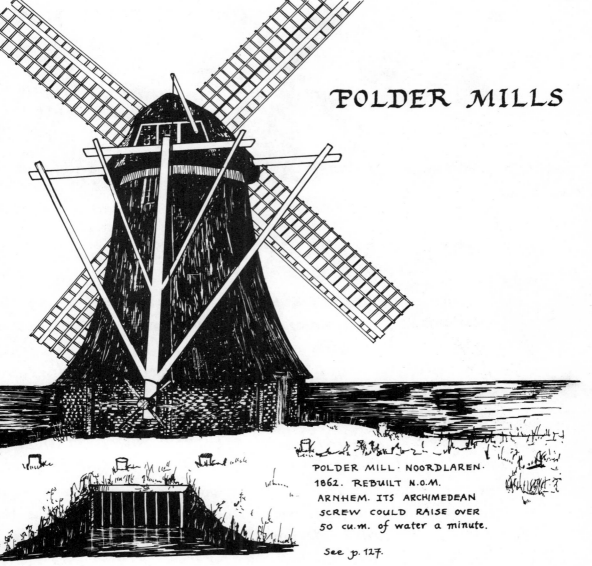

POLDER MILL · NOORDLAREN.
1862. REBUILT N.O.M.
ARNHEM. ITS ARCHIMEDEAN
SCREW COULD RAISE OVER
50 cu.m. of water a minute.

See p. 127.

The polder lands of Holland were reclaimed by mills like this one. The design dates from the mid C17 when scoop wheels began to be replaced by Archimedean screws. The gear train in the polder mill is the same as the spider mill's ~ p.126. Fifty-one polder mills were needed to create the Schermerpolder (1631~35). Together they discharged some 1000 cu. meters of water a minute. Polder mills were often arranged in groups or gangs ~ like the sixteen mills at Kinderdijk ~ Zuid Holland.

MEADOW MILLS

In North Holland there has always been a need to keep the meadowlands well watered. There the fields are divided into narrow strips with a water channel between. The lift needed to raise the water is slight, and the Dutch millwrights developed a suitable sized mill ~ about 10ft. high ~ called a meadow mill. There are two Dutch words for this type of mill ~ WEIDMOLENTE & AANBRENGERTJE. Later examples of these mills have wooden shutters which fit the sail frames. The task of rigging the sails can be performed by youths. A large vane at the rear keeps the sails towards the wind. In operation the sails turn a vertical shaft which has three or four blades at its base. These blades are below the water level. As the shaft rotates the blades stir the water which rises through centrifugal force ~ like stirring a cup of tea. The raised water flows over the sill into the drainage ditch.

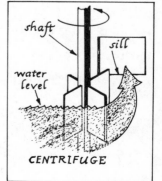

shaft

sill

water level

CENTRIFUGE

TJASKERS

ne of the most curious mills is he TJASKER which was used mostly in Friesland. The mill has n inclined shaft, which embodies n Archimedean screw, with sails t its upper end. As the common weeps rotate the shaft the crew raises the water. t then spills into he outlet canal. ratchet n the haft

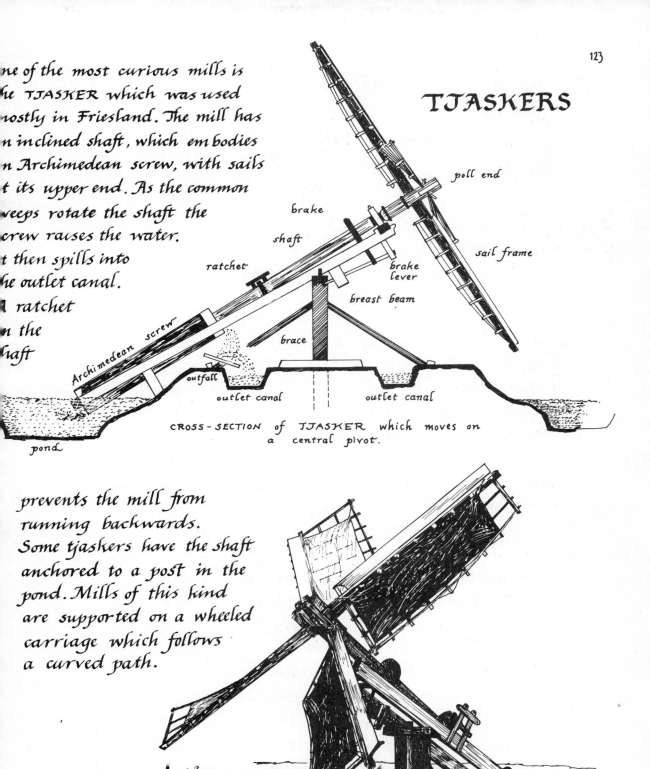

CROSS-SECTION of TJASKER which moves on a central pivot.

prevents the mill from running backwards. Some tjaskers have the shaft anchored to a post in the pond. Mills of this kind are supported on a wheeled carriage which follows a curved path.

THE WIP MILL

Perhaps the most attractive of the Dutch mills is the WIP MILL with its pyramidal base and well proportioned body that carries the sails. The wip mill usually functioned as a drainage mill but some were used for corn. Wip mills which look like post mills are technically hollow post mills. This means that the central post which supports the mill body is hollow, & the drive shaft linking the sail-power to the scoop wheel passes through it.

The pyramid base of the wip mill serves the same purpose as the post mill's trestle. In the diagram two lantern gears are shown on the vertical shaft. This style of gear seems to have remained in use in Holland long after English millwrights had discarded it. As the common sails turned, the brake wheel & its wallower moved in unison. The lower lantern wheel and the under wheel worked the scoop wheel at a slower pace than the sails. The size of the scoop wheel is in proportion to the span of the sails. When the WIP MILL was evolved in the early C15 its inventors seem to have mastered the mathematics by intuition. A windlass at the foot of the steps was used to luff the sails. The brake can also be operated from the same position.

To protect it from the weather thatch covered the base. As this part of the mill was also used as a dwelling windows were needed. There are always two doors in the mill base. These are placed at right angles to the scoop wheel. This arrangement allowed its occupants a safe exit whichever way the sails faced. When the mill was working the door facing the sails was kept locked.

One feature of these mills is the highly colourful manner of their decoration. They make distinctive landmarks in the polder.

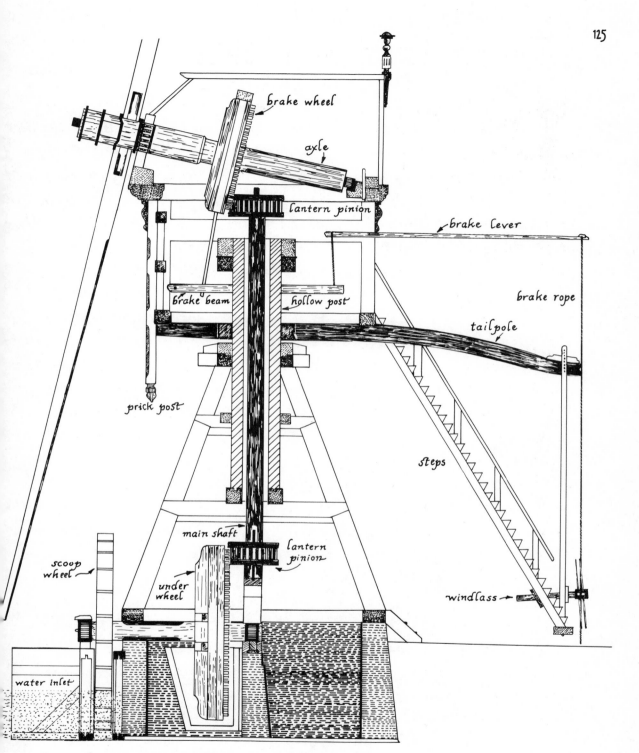

brake wheel

axle

lantern pinion

brake lever

brake beam

hollow post

brake rope

tailpole

prick post

steps

main shaft

lantern pinion

scoop wheel

under wheel

windlass

water inlet

CROSS SECTIONAL VIEW OF WIP MILL

SPIDER MILLS

In Friesland the wip mill has a smaller relation which the Dutch call a spinnekopmolen ~ spider mill. It too has a hollow post and its features are a duplication of the wip mill. The diagram opposite shows the principal elements of the spinnekop. These mills stood about twenty feet high and had a sail span of about thirty-five feet. Most spider mills had a square pyramid base, but hexagonal plans were also used. The base was boarded or protected with tiles.

Spider mills used scoop wheels or archimedean screws. Although the spinnekop was not as efficient as the wip mill it served the needs of farmers well enough. Some versions of the spider mill had a large vane at the rear like the meadow mill on p. 122. Spider mills were often located in isolated places.

FRIESLAND SPIDER MILL with base protected by tiles.

A considerable number of spider mills remain in Friesland. Some of them date from the early 1800s. Like most Dutch mills they have a personal name which helps to make them more interesting even though we may not know why a particular cognomen was chosen. Among the spider mills we can find "De Bird", "De Ikkers" and the grand-sounding "De Princehofmolen".

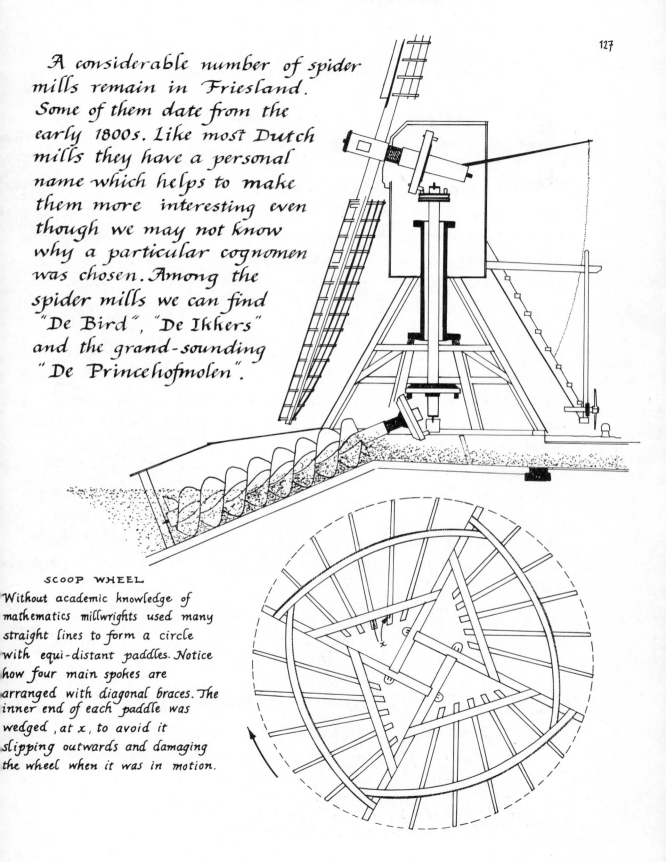

SCOOP WHEEL

Without academic knowledge of mathematics millwrights used many straight lines to form a circle with equi-distant paddles. Notice how four main spokes are arranged with diagonal braces. The inner end of each paddle was wedged, at x, to avoid it slipping outwards and damaging the wheel when it was in motion.

BERNEY ARMS

ngland's most inaccessible mill must be the very fine tower which presides over the landscape about the River Yare ~ just above Breydon Water. You can reach it by train ~ Berney Arms station, by boat or walking across the marshes from Halvergate some three miles distant, but not by road. There is a dramatic elegance about its setting. It is best appreciated by watching the scudding alto-cumulus crossing the tower's silhouette at sunset; or the mares' tails that so often provide the backdrop for its form against that immensely blue Norfolk sky.

Since the 1860s it has stood against the elements &, until 1948, fulfilled its twofold purpose of draining the adjacent marsh and grinding clinker for cement.

Its brick tower is about 70 feet high, and the detached scoop wheel has a diameter of 24 feet. This wheel driven by the horizontal shaft ~h~ and a spur gear which works a cog ring ~p.53. Inside a pair of edge-runner stones remain ~p.43. The mill, now in the care of the Dept. of the Environment, is open daily from April to September.

THE ANNULAR SAIL

As the C19. progressed engineers developed less costly machines to lift water. These dispensed with timber and relied upon cast iron gearing with a framework of iron and steel. The power generated by these skeletal engines was enhanced by the use of the annular sail, patented 1855, with its multiple blades which increased the sail's effective area. The introduction of this form of sail has been attributed to Mr. Ruffle who had such a sail fitted to his tower corn mill at Haverhill, Suffolk (c.1861). At about the same time a similar sail was added to the smock mill at Boxford in the same county.

The apparatus shown here was advertised in the 1890s by John Wallis Titt, a water engineer, of Warminster, Wilts. The advantage of this type of machine was its relative cheapness when compared with a brick tower. During the Victorian era the application of this type of gear was not restricted to Britain. Its portability allowed it to reach into the New World as well as Africa & Australasia.

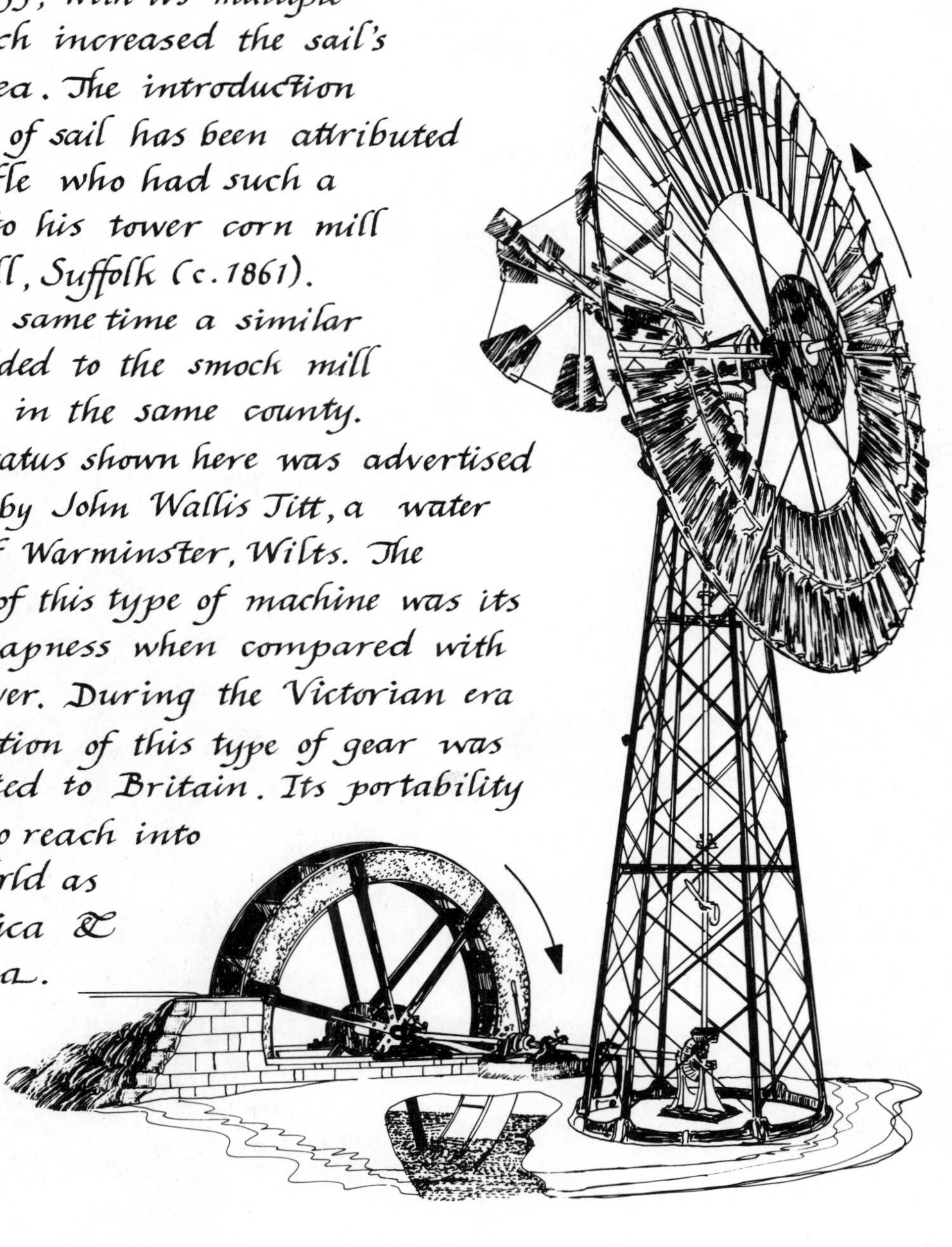

CHESTERTON, WARWICKSHIRE.
This fine mill is the most unusual one in England. Its stone tower is pierced by six elegant arches. Built in 1632 the machinery is still in good order.
There is an 8ft. brakewheel with compass arms which drives a substantial lantern pinion placed at the top of the vertical shaft. Its domed cap is turned by an internal hand winch.
The mill is open one week-end each alternate year. Then steps are erected to provide access to the interior. The mill is owned by the COUNTY COUNCIL.

STRACEY ARMS, R. BURE, NORFOLK.
A tower wind pump with a galleried boat-shaped cap, and a fanstage of generous proportions. The Y-wheel can be clearly seen below it. Note how the once shuttered sails are twisted to catch the wind.

This mill is open during the summer season.

EASTBRIDGE WINDPUMP,

STOWMARKET, SUFFOLK.
A view of the fanstage.
The chain x works
the wheel controlling
the blades in the sails
~ see p. 105.
Wheels of this type are
called Y-wheels. Note the
Y-shaped grooves around
the circumference to
carry the chain.
A sectional drawing
of the mill is shown
on p. 119.

SUTTON MILL, STALHAM, NORFOLK.
This is England's tallest mill. The
tower (rebuilt in 1857) is 80ft. high,
and the cap 12ft. Its sails-when
restored will have a span of 73ft.
There are eight floors & four pairs
of stones. The drawing shows the mill
in the 1930s. It ceased work in 1940
when it was damaged by lightning.
Christopher Nunn, its present owner,
is gradually restoring the mill.
Visitors can also admire his extensive
collection of country bygones.

SUGAR MILLS ~ BARBADOS

In the mid C17 the sugar producers of Barbados began to use the power of the Tradewinds to crush their cane. The crushing mechanism was the same as that used in the cattle mills ~ p.34. The typical sugar mill on the island has a Dutch look with its characteristic long pole ~ a tail race ~ that projects from the rear of the cap and reaches the ground. To make it easier to move the tailrace a cart wheel was often fixed to the pole's lower end. This was a common feature on many English post mills. One reason for the very long tailrace was the need for it to clear the roof of the crushing house when the cap was turned. The internal machinery of a sugar mill was simple.

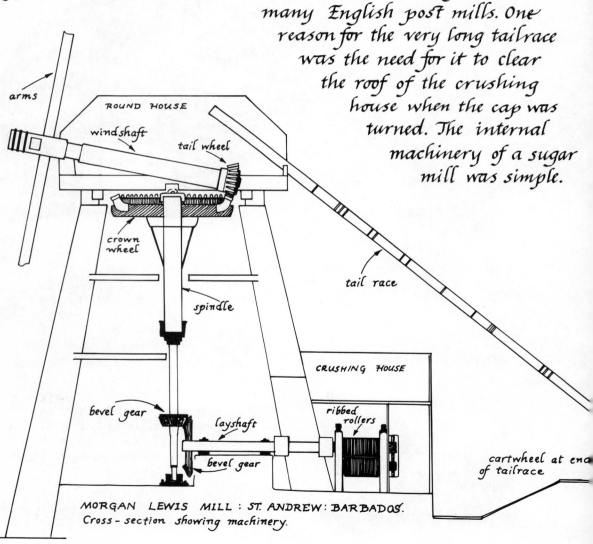

arms

ROUND HOUSE

windshaft

tail wheel

crown wheel

spindle

tail race

bevel gear

layshaft

CRUSHING HOUSE

ribbed rollers

bevel gear

cartwheel at end of tailrace

MORGAN LEWIS MILL : ST. ANDREW : BARBADOS.
Cross-section showing machinery.

Common sails turned the windshaft ; and a bevel gear at its tail end moved the crownwheel at the top of the spindle. This passed through the intermediate floors to the ground. Another bevel gear then worked the layshaft and the ribbed rollers.

Most sugar farms were small and the windmill was placed on the highest part of the site. It was then possible to pipe the juice downhill to the boiling house. Crushed cane ~ bagasse ~ was dried and used as fuel to boil the juice. Farms sometimes had more than one mill. Before the end of the C17 there were as many windmills as cattle mills in Barbados. About 500 windmills operated in the C18 and cattle mills became fewer. Steam power was introduced in the 1840s and its use led to the demise of the windmill. The last working mill was at Colleton St. Peter. It ceased work in 1944.

Given sufficient wind and good quality cane a mill could process some 200 tons of cane each week. This would result in more than 5000 gallons, 22,730 litres, of syrup. Sugar production needed a lot of labour. About 20 workers attended the mill and boiled the syrup. A similar number were employed in the field for cutting and carting.

The Morgan Lewis mill at St. Andrew is now owned by the Barbados National Trust. It has been restored & is open to visitors.

SAWMILLS

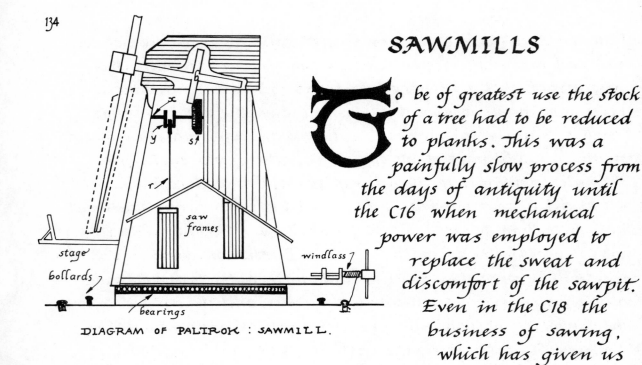

DIAGRAM OF PALTROK : SAWMILL.

To be of greatest use the stock of a tree had to be reduced to planks. This was a painfully slow process from the days of antiquity until the C16 when mechanical power was employed to replace the sweat and discomfort of the sawpit. Even in the C18 the business of sawing, which has given us the surname SAWYER, was often a task for itinerant men whose wages of so much a foot of sawing had to be supplemented by free ale. The first sawmill seems to have appeared in Holland ~1592~ when Cornelis Corneliz, of Uitgeest, patented the idea. His mill was named "The Damsel" and four years after its construction it was moved by water to Zaandam, a distance of about 8 miles. From this mill the design of the PALTROK is derived.

In the diagram above the principal features of the paltrok are shown. The mill body rested upon a masonry base and it was turned by a windlass. To make the task easier a ring of roller bearings is positioned at the base of the tower. Sails could be adjusted from the stage at the front of the mill. The machinery of the paltrok is much simpler than a corn mill. A brake wheel ~b~ turns with the windshaft and operates a secondary gear ~s~ that rotates a horizontal shaft ~x. Two cranks ~y~ on this shaft pull the saw frames up and down via connecting rods ~r. The paltrok has a

PALTROK
MILL.
ZAANDAM

distinctive shape to allow the
maximum amount of room on
the sawing floor. Two wings
project from the tower to give
shelter to the working area.
A crane at one end lifted
the tree trunks. As the
sawing progressed the
timber was slowly moved
towards the saw frames. One
of the great advantages of
a sawmill was its
capacity to saw several
planks at the same
time.

SMOCK TYPE
SAWMILL

Many Dutch sawmills were
smock mills. This form of
tower could be placed upon
a range of sheds and the
wing extensions on the paltrok
were not needed. Sawmills
were placed next to a
waterway so that timber
could easily be delivered.
Smock sawmills had
a slipway so that trees
could be winched into
the mill.

ALTERNATIVE POWER

After many years of neglect the free power of the wind is now being re-valued. In an age when pollution is rife, and man's exploitation of non-renewable energy has never been more pronounced, some people are beginning to re-assess the total worth of our planet's wind energy. The Cretan-style mill with its distinctive jib sails can still be seen on that island. A version of it has been devised at the Centre for Alternative Technology to generate electricity. Its sails are mounted on a simple timber tower, and a broad vane at the rear keeps the sails facing the wind. Detailed drawings are available from C.A.T. for those who wish to build their own wind engine. With a sail diameter of 12ft. it can work a lathe, pump, or provide up to 700 watts of power. The Cretan mill at Machynlleth is made from simple, and often secondhand, components. Its brake gear is derived from a car.

CRETAN MILL · CENTRE FOR ALTERNATIVE
TECHNOLOGY · MACHYNLLETH · POWYS

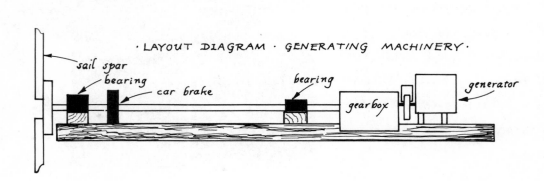

· LAYOUT DIAGRAM · GENERATING MACHINERY ·

sail spar

bearing

car brake

bearing

gearbox

generator

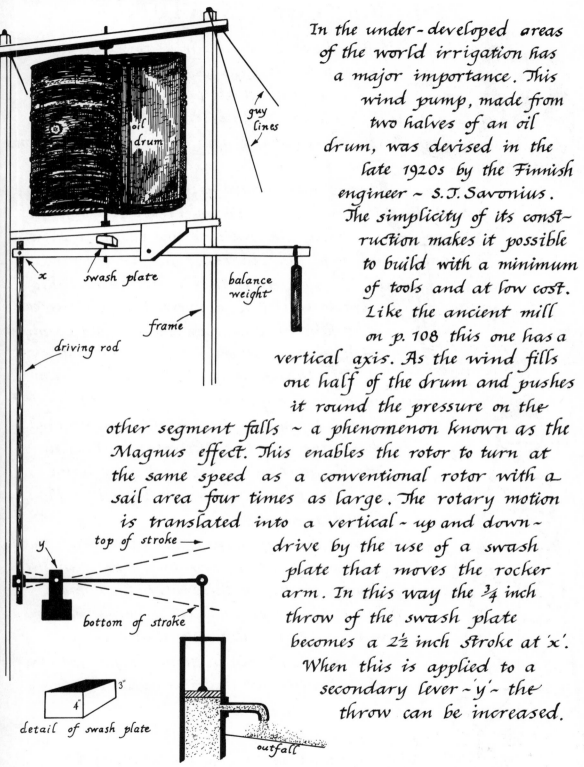

guy lines

oil drum

x

swash plate

balance weight

frame

driving rod

top of stroke →

y

bottom of stroke

3"
4"

detail of swash plate

outfall

In the under-developed areas of the world irrigation has a major importance. This wind pump, made from two halves of an oil drum, was devised in the late 1920s by the Finnish engineer ~ S. J. Savonius. The simplicity of its construction makes it possible to build with a minimum of tools and at low cost. Like the ancient mill on p. 108 this one has a vertical axis. As the wind fills one half of the drum and pushes it round the pressure on the other segment falls ~ a phenomenon known as the Magnus effect. This enables the rotor to turn at the same speed as a conventional rotor with a sail area four times as large. The rotary motion is translated into a vertical ~ up and down ~ drive by the use of a swash plate that moves the rocker arm. In this way the ¾ inch throw of the swash plate becomes a 2½ inch stroke at 'x'. When this is applied to a secondary lever ~ 'y' ~ the throw can be increased.

A working drawing of this mill can be obtained from C.A.T.

THE AEROGENERATOR

Many years before the oil crisis made us consider our power resources, wind engines had been used to pump water and generate electricity in distant places like S. Africa, N. America & Australia.

National and private companies are now developing new forms of wind generator. In the British Isles the Central Electricity Generating Board has several experimental machines. Two of them are located at Carmarthen Bay and at Burgar Hill, Orkney. The latter engine has a 262 ft. high tower and 65 ft. dia. blades. It has an output of 700,000 kwh. The success of this wind engine has persuaded the Government to approve the construction of a further generator three times as big with an output of 9 million kwh. p.a.

The location of a wind generator is a matter of importance. Experimental projects are all sited in places with good levels of wind throughout the year. The Netherlands Energy Research Foundation's horizontal axis wind turbine HAWT is situated at Petten on the North Holland coast.

Ordnance Survey maps of Britain still record the positions of the old wind pumps with this symbol ✤. Although these engines, in more affluent days, were neglected many have now been refurbished. Their stature in the landscape made them useful points of reference for mapmakers.

CONTROL ROOM

C.E.G.B. AEROGENERATOR, CARMARTHEN.

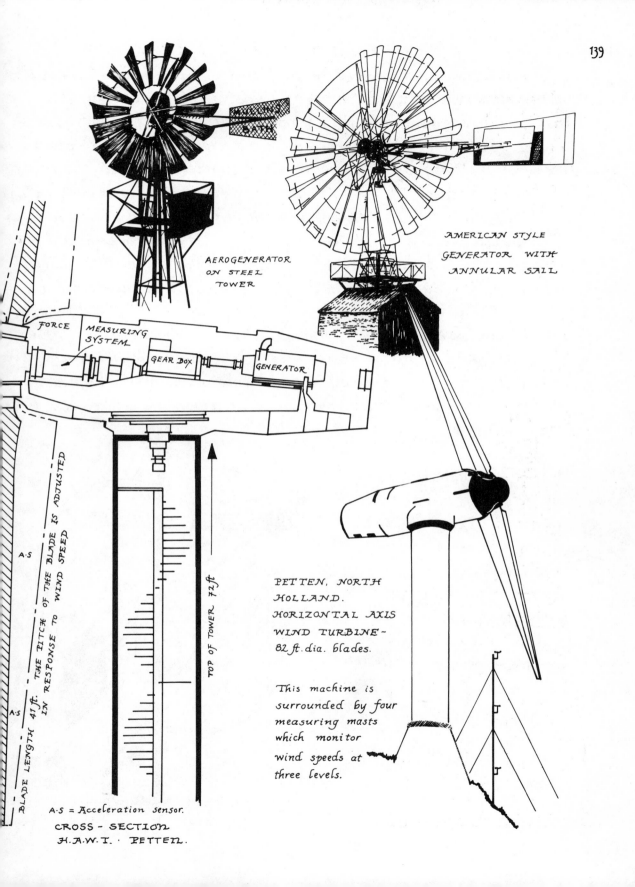

AEROGENERATOR
ON STEEL
TOWER

AMERICAN STYLE
GENERATOR WITH
ANNULAR SAIL

FORCE MEASURING SYSTEM

GEAR BOX

GENERATOR

BLADE LENGTH 41 ft. — THE PITCH OF THE BLADE IS ADJUSTED IN RESPONSE TO WIND SPEED

A·S

A·S

A·S = Acceleration sensor.

TOP OF TOWER 72 ft.

PETTEN, NORTH HOLLAND. HORIZONTAL AXIS WIND TURBINE - 82 ft. dia. blades.

This machine is surrounded by four measuring masts which monitor wind speeds at three levels.

CROSS - SECTION
H.A.W.T. · PETTEN.

Engineers, particularly in Holland, have developed several forms of aerogenerator. The twin mast turbines shown here can produce power ranging from 30~120 kws. Catwalks give access to the working parts. As the wind changes direction the arm, resting on a roller track mounted around the mast, can turn to face the wind. The design shown is manufactured by Kaal van der Linden, Oss, Netherlands. This company also builds multi-turbine systems ~ with six generators on a mast ~ to suit specific needs. The use of computers allows an accurate analysis of cost-effectiveness to be made for a given site and wind conditions. There are more than twenty Dutch companies which manufacture aerogenerators. About half of them produce machines with lower output levels suited to the needs of modest consumers. One of the biggest problems is storing the power generated for use at a later time when wind energy is not available.

Not all generators are factory made. This unusual example with a cone-shaped rotor was devised by a Sussex farmer to provide lighting for a cow shed. It can produce about 3 kw and will work in a light breeze.

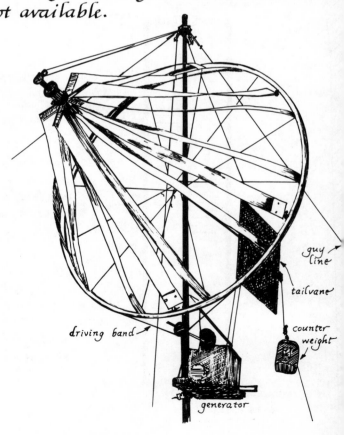

guy line

tailvane

driving band

counter weight

generator

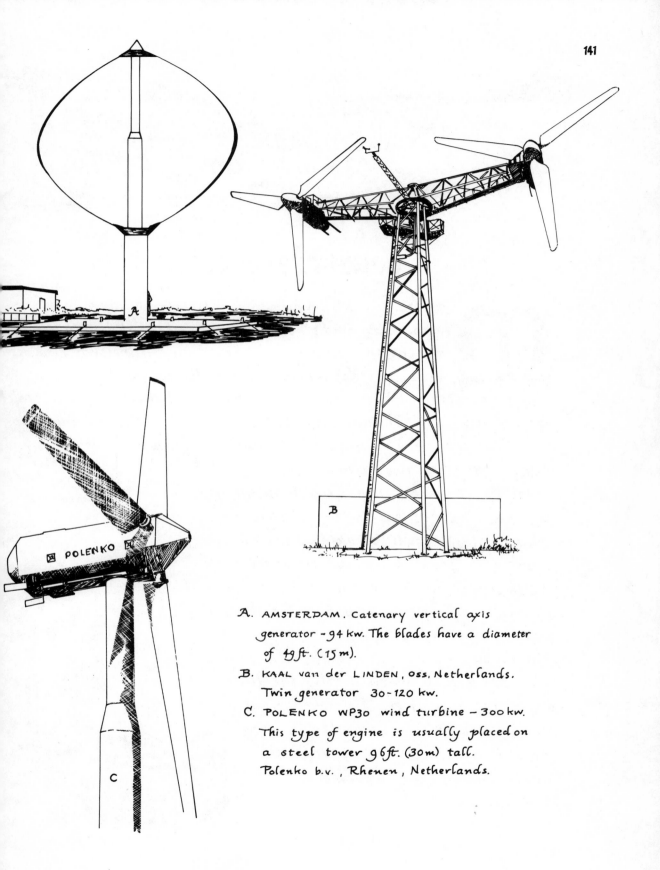

A. AMSTERDAM. Catenary vertical axis
   generator - 94 kw. The blades have a diameter
   of 49 ft. (15 m).

B. KAAL van der LINDEN, Oss, Netherlands.
   Twin generator 30-120 kw.

C. POLENKO WP30 wind turbine - 300 kw.
   This type of engine is usually placed on
   a steel tower 96 ft. (30 m) tall.
   Polenko b.v., Rhenen, Netherlands.

pintel

COUNTER WEIGHTS

Man has made much use of weights to save himself labour. A weight can be pushed and left to apply its energy in a given direction. In the two Dutch examples drawn here a small tree with part of its root structure provides the counter balance. The gate, which in England would be called a heave gate, rests on a post that has a vertical iron pin ~ a pintel. This acts as a hinge.
A forked timber provides the resting post.
Three braces are suspended from the gate bar
to make fixings for the two irregular
planks that form the gate's substructure.
These members are joined together with
wooden dowels. When the gate was opened
the overhanging bar helped to lighten the
load.

The bucket, hooked to the stick, is dipped
into the well & when it fills the stick
is jerked upwards. Then the counter-
balance acts to hasten its journey.
An old cartwheel was placed
across the well to prevent
accidents.

GRAVITY

The principle of gravity had been hidden from mankind since the Creation until its secrets were unlocked by Isaac Newton (1642 - 1727). There are few people who do not know the story about the apple that inspired his investigations. At Woolsthorpe Manor, Lincs. (National Trust) visitors can still see the garden where he sat - 1665 - when he returned from Cambridge to escape the plague. Today's apple trees are descended from those Isaac Newton knew.

Long before Newton's acute mind focussed upon the scientific aspects of gravity men had put it to work. The first machines driven by weights were those intricate but robust clocks made by blacksmiths from the C14 onwards.

Man also put weights to work for him in other ways. On heavy ground weights were often added to the plough beam. They were used to close gates, provide counterbalances and to add force to the beam drills once found in craftsmen's workshops.

One of the most unusual examples of working gravity can be seen on the steep cliffs of North Devon. There the weight of water in a tank has supplied the power to operate a cliff railway for almost a century. The advantage of gravity is that it is almost free.

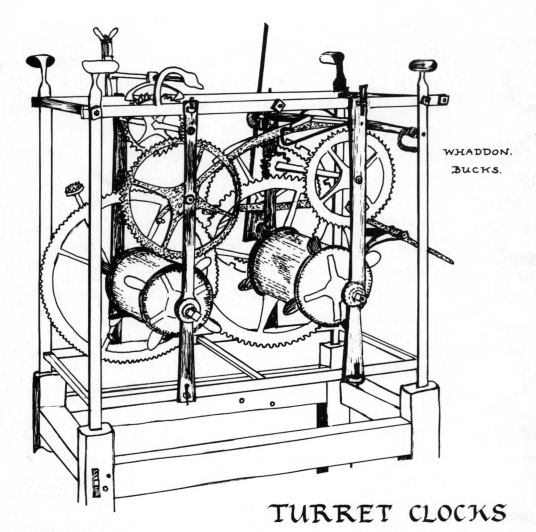

WHADDON.
BUCKS.

TURRET CLOCKS

The mysteries of measuring time in a mechanical manner eluded man until the Middle Ages when blacksmiths began making their weight-driven masterpieces that were the wonder of their age. Technology was significantly advanced by these early and mostly anonymous horologists. Their work was helped in c.1657 when Christian HUYGENS introduced the pendulum. England's oldest working clock is probably the splendid machine c.1386 at Salisbury Cathedral. An C18 version still works at Uffington, Vale of the White Horse. The first clocks had no hands, & they just struck the hours. They were literally "wound up" when their barrels were turned to pull up the weights attached to them. It was the

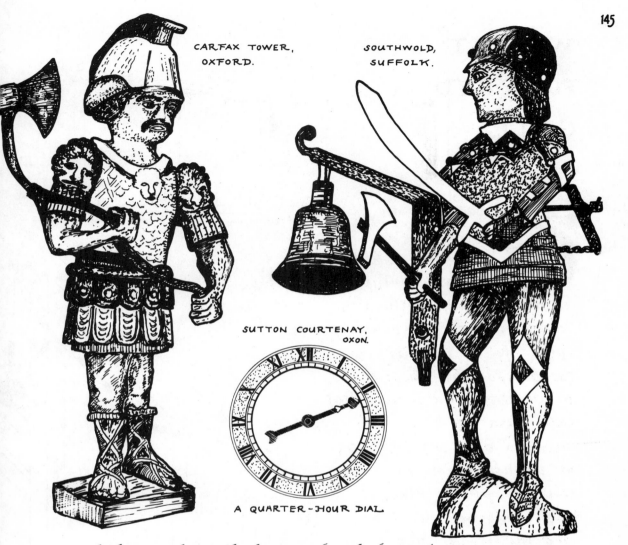

CARFAX TOWER, OXFORD.

SOUTHWOLD, SUFFOLK.

SUTTON COURTENAY, OXON.

A QUARTER-HOUR DIAL

action of the weights which gave the clock its driving power. Striking clocks had two sets of gears – called the 'going' & 'striking' train. Mediaeval engineers also invented animated figures – JACKS – which were operated by the clock's works. Unknowingly they had ushered in the age of automation. Clockmakers also made use of the first power-take-off devices when they added a hand to show the hour. The first clock faces had just one hand. The space between each hour was sometimes divided into four to show the quarters. It was the introduction of the minute hand, c.1690, which gave us our modern clockface.

WEIGHT POWERED GATE

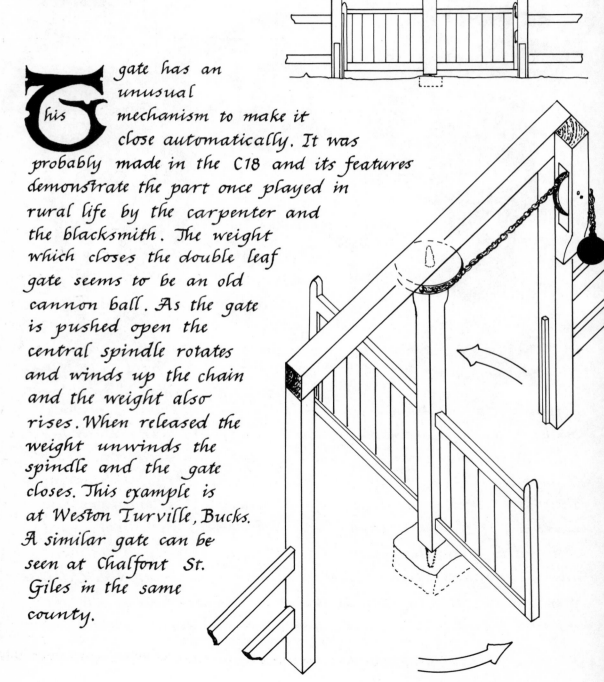

This gate has an unusual mechanism to make it close automatically. It was probably made in the C18 and its features demonstrate the part once played in rural life by the carpenter and the blacksmith. The weight which closes the double leaf gate seems to be an old cannon ball. As the gate is pushed open the central spindle rotates and winds up the chain and the weight also rises. When released the weight unwinds the spindle and the gate closes. This example is at Weston Turville, Bucks. A similar gate can be seen at Chalfont St. Giles in the same county.

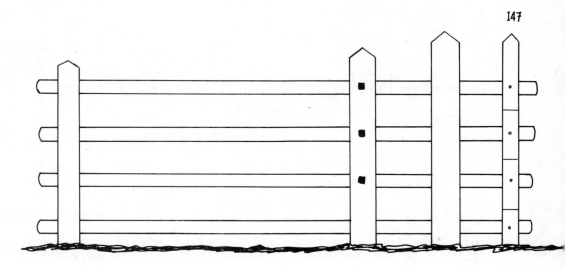

CLAPPER or TUMBLE STILE

Gravity was put to work in many ways. This drawing shows a rather unusual stile that makes use of a counter weight. At first sight the stile looks like a fence with irregularly spaced posts. The lower drawing shows what happened to the rails when the top rail, on the left hand side was pushed downwards. Examples have been recorded in Berkshire, Cambridgeshire, Derbyshire & Sussex. Those known by the author to survive can be seen at Hungerford, parish church, Berks ; Sissinghurst Castle, Kent ; Charlcote Park, Warwicks.

DRILLS

From prehistoric times the bow drill was used for making holes. This tool has a vertical spindle and a horizontal 'bow' to which the string is fixed. As the bow was moved up and down the spindle rotated in alternate directions. Drill bits used in a bow drill had to cut in clockwise and anti-clockwise directions. The principle was also used to make fire, the vertical hardwood spindle rotating and creating heat from its friction with a piece of softwood.

Drilling metal required a constant & significant pressure on the drill bit. One method of providing the necessary force was to use a screw to drive the drill into the object worked. The drill clamp had to be held in a vice and the drill was fixed into the clamp in the manner shown. As the bit cut into the surface of the workpiece the screw adjuster was gradually turned. This forced the drill and the bit downwards with some considerable pressure. Although slow this method of drilling was effective. One advantage of the drill clamp was its portability.

VICE

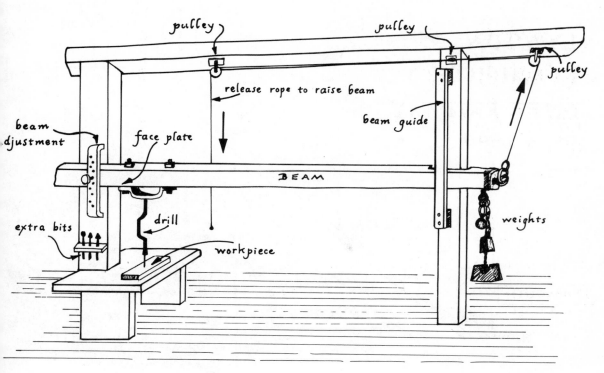

To create a steady pressure on a drill required a good deal of energy. In a workshop, beam drills were often used. These employed the weight of a substantial wooden beam to push the drill downwards. To add to the power of the beam heavy weights were attached to its end in the way shown in the drawing. At the conclusion of drilling the beam had to be lifted to free the drill, and release the workpiece. Pressure could be released by raising the beam with the pulley rope that was suspended near the bench. Workpieces varied in size & it was necessary for the drill to be adjustable. This alteration was made by moving the pin in the adjuster plate which had many settings. The thrust of the beam was transferred to the drill via the face plate bolted to the underside of the beam. As beam drills were home-made, they never conformed to a standard design.

LYNTON & LYNMOUTH CLIFF RAILWAY ~ 1890 ~

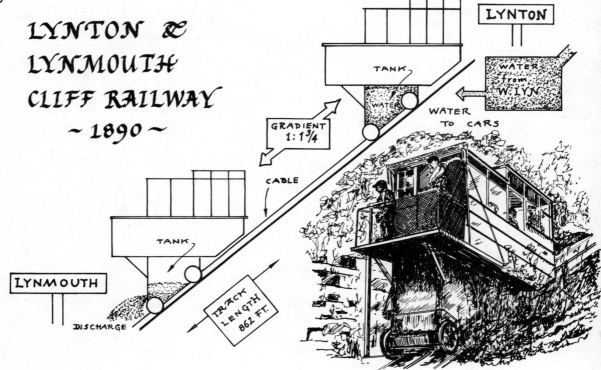

LYNTON

WATER from W. LYN

TANK

WATER

WATER TO CARS

GRADIENT 1:1¾

CABLE

TANK

LYNMOUTH

DISCHARGE

TRACK LENGTH 862 FT.

Perhaps the most unusual example of a counterbalance to be seen in England is the spectacular cliff railway at Lynton, Devon. Opened in 1890 & designed by Bob Jones, whose grandson is the present Engineer, the railway gives its passengers some breathtaking views. There are two cars. Gravity provides the motive power. The water supply comes from the West Lyn River. Below each car is a tank with a capacity of 700 galls. With a full tank and passengers a car weighs 10 tons. When a car at Lynton has filled its tank a signal is sent to the driver at Lynmouth. The bottom car discharges its water. When the hydraulic brakes are released the heavier vehicle can start its descent. As the upper car moves down the lighter car is pulled upwards by the connecting cable. Before cars could climb the road to Lynton they were taken up the railway at 7s.6d. a time. Coal, sand, granite & oil have also been carried by Lynton's quiet, pollution free and romantic railroad. The water-powered lift on the cliff at Saltburn, Cleveland has been in use since 1884.

SASH WINDOWS

Every day thousands of people open or close their sash lights. This distinctive feature of English architecture ~ some say the design came from Holland ~ adds particular grace to our Georgian heritage. Few stop to consider how a sash mechanism works.

This diagram shows how moving a sash causes weights to rise. When the window is closed both weights are at x. The combined pull of the weights balances the mass of the window and so it stays in any selected position.

The balance could be upset if a sash cord snapped. Then the secret door at the bottom of the pulley stile has to be opened to remove the weight; and the appropriate frame taken out. This is usually a job for the glazier or carpenter.

Vertical sashes were first used in great houses in the C17. By Queen Victoria's reign they began to find a place in more modest terraces in our industrial towns, and by the end of the century they were ubiquitous.

SASH WINDOW

pulley stile door

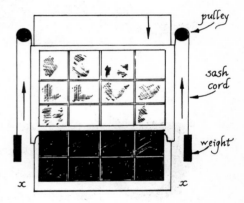

pulley

sash cord

weight

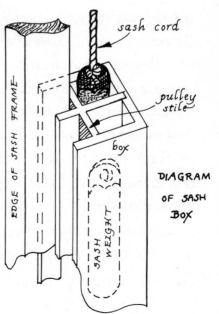

sash cord

pulley stile

EDGE OF SASH FRAME

box

SASH WEIGHT

DIAGRAM OF SASH BOX

CREDO

MICHELHAM PRIORY,
SUSSEX.

There is a twofold significance in preserving the tools and the workplaces of the past. They can in a direct manner allow our senses to experience the size, sound and motion of an earlier technology. Their second function is to sustain the skills and knowledge which can still serve the needs of man. Environmental disasters of the Three Mile Island or Windscale kind may help us to revalue and re-instate some of the older methods that C20 western man has brushed aside, but the third world still employs. The re-emergence of windpower and its use to generate electricity, in places as far apart as Orkney and Australia, is a positive sign. Fossil fuel is finite but wind and water power are re-newable. The study of the power potential of tidal energy is another laudable development.

In this century man has tried to go too fast down several technical highways - to discover that they lead nowhere.

The new awareness of the need to husband our finite resources has produced enterprises like the Centre for Alternative Technology · p.154. There old ideas are combined with new techniques to create an appropriate technology. That is one which is capable of serving the needs of all humanity indefinitely, and not just those domiciled north of the equator. A classic account of these ideas can be found in Dr. E.F. Schumacher's 'Small is Beautiful'.

The western world must be prepared to accept limits to its level of consumption. Without that pre-requisite we shall, at best, condemn the other half of mankind to forgo even the rudiments of a meaningful life. One man's conspicuous consumption is probably another's chance of survival.

There is still time for us to choose which path to follow, & to find ways of conserving our non-renewable energy. But which future will we adopt?

MEA CULPA!

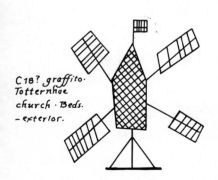

C18? graffito.
Totternhoe
church · Beds.
- exterior.

POST MILL · AVONCROFT MUSEUM · BROMSGROVE · WORCS.

PLACES TO VISIT

This is a selective list of fascinating, & highly recommended, places to visit. The figures at the end of each entry indicate the type of power sources to be found: viz. 1 man, 2 animal, 3 wind, 4 water, 5 gravity. Detailed information about many other water & windmills open to visitors may be found in the author's 'Discovering Watermills' - 4th. edit. 1984; & 'Disc. Windmills' - 6th. edn. 1984: Shire Publications.

Abbeydale Industrial Hamlet, Sheffield. (0742) 367731   4

Alford Mill, Alford, Lincs. (0652) 648382   3

Armley Mills, Canal Rd., Leeds. (0532) 637861   1,4

Avoncroft Mus. of Buildings, Stoke Heath, Bromsgrove, Worcs.
(0527) 31363   1,3

Beamish: North of England Open Air Museum, Co. Durham.
(0207) 31811   1,2

Berney Arms Mill, R. Yare, Norf. (0493) 700605   3

C.A.T.: Nat. Centre for Alternative Technology, Machynelleth.
(0654) 2400   1,3,4

Claverton: The American Museum, Claverton Manor, Bath.
(0225) 60503   1

Claverton Pump House, Ferry Lane, Claverton. (0272) 712939   4

Cogges: Manor Farm, Cogges, Witney, Oxon. (0993) 72602   1,2

Dunham Massey, Altrincham, Cheshire. (061 941) 1025   4

Eling Tide Mill, Southampton. (0703) 869575   4

Gunton Park Sawmill, Norf. (0603) 611122 Ext. 481   4

Harwich Treadwheel Crane.   1

Heatherslaw Mill, Ford, Northumb. (089-082) 291   4

Heckington Mill, Lincs. (0529) 60241   3

Highland Folk Museum, Kingussie. (05402) 307   4

Hunday: Nat. Tractor Museum, Stocksfield, Northumb.
(061) 842553   1,2,4

Laxey: Lady Isabella Wheel, I.o.M. (0624) 74323   4

Lynton Cliff Railway, Lynton, Devon. ... 5
Mary Arden's House, Wilmcote, Warw. (0789) 4016 ... 1,2
Michelham Priory, Upper Dicker, E. Sussex. (0323) 844224 ... 2,4
Morwellham Quay, Tavistock, Devon. (0822) 832766 ... 4
Museum of Cider, Hereford. (0432) 54207 ... 1,2
Museum of East Anglian Life, Stowmarket, Suff. (0449) 61229 ... 3,4
Museum of English Rural Life, Univ. of Reading. (0734) 875123 ... 1,2
Muncaster Mill, Ravenglass, Cumbria. (06577) 232 ... 4
Ponterwyd: Llywernog Silver-Lead Mine, Dyfed. (097085) 620 ... 4
Quarry Bank Mill, Styal, Cheshire. (0625) 527468 ... 4
Redditch: Forge Mill Mus. (0527) 62509 ... 4
Ryedale Folk Museum, Hutton-le-Hole, N. Yks. (075-15) 367 ... 1,2
Saltburn Cliff Railway, Saltburn, Cleveland. ... 5
Sticklepath: Finch Foundry Trust, Okehampton, Devon.
(083784) 286 ... 1,4
Stoke Bruerne: Waterways Museum, Nhants. (0604) 862229 ... 1,2
Sutton Mill, Stalham, Norf. (0692) 81195 ... 3
Weald & Downland Open Air Museum, Singleton, Sussex.
(024 363) 348 ... 2,3,4
Wellbrook Beetling Mill, Co. Tyrone. (039686) 204 ... 4
West Blatchington Mill, Brighton. (0273) 734476 ... 3
Wheal Martyn, Carthew, St. Austell. Corn. (0726) 850362 ... 4
Wicken Fen, Cambs. (0353) 720274 ... 3
Wilton Mill, Gt. Bedwyn, Wilts. (0672) 870268 ... 3
Woodbridge Tide Mill, Suff. (03943) 2548 ... 4
Wookey Hole, Wells, Som. (0749) 72243 ... 4
Worsborough Mill Museum, Barnsley. (0226) 203961 ... 4
Wortley Top Forge, Sheffield. (0742) 343479 ... 4
HOLLAND: The Netherlands Open Air Museum, Schelmseweg 89, 6816 SJ, Arnhem.

Some of the places listed restrict opening to the summer months. Readers are advised to check before making a visit.

BIBLIOGRAPHY

Beeson, C.F.C.	English Church Clocks 1200~1850	Phillimore
Birdsall, D.+ Cipolla, C	The Technology of Man	Wildwood
Booth, D.T.N.	Warwickshire Watermills	MIDLAND MILLS GROUP
Brunnarius, M.	Windmills of Sussex	Phillimore
Darrow, K + Rich, P.	Appropriate Technology Source Book	U.S.A. †
Dolman, Peter	Windmills in Suffolk	SUFFOLK MILLS GROUP
Foreman, Wilfred	Oxfordshire Mills	Phillimore
Flint, Brian	Suffolk Windmills	Boydell
Freese, Stanley	Windmills & Millwrighting	C.U.P.
Jesperson, A.	The Lady Isabella Waterwheel	E.C. KNEALE. Laxey, I.O.M.
Lawson, D.+ Jackson, H.	Hand to the Plough	Ashgrove
Major, J.K. + Watts, M.	Victorian & Edwardian Windmills & Watermills	Batsford
Major, J.K.	Animal Powered Engines	Batsford
Reynolds, John	Windmills & Watermills	H. Evelyn
Salaman, R.A.	Dictionary of Tools	Allen +Unwi
Seaby, W.+ Smith, A.C.	Windmills in Staffordshire	STEVENAGE MUSEUM
Smith, A.C.	Corn Windmills in Norfolk	"
	Windmills in Cambridgeshire	"
Stokhuyzen, F.	The Dutch Windmill	Van Dishoeck
Taylor, R.H.	Alternative Energy Sources	Hilger
Vince, John	Old Farms	J.Murray
	Illustrated Hist. of Carts & Wagons	Spur
	Farms and Farming	I. Allan
	Village Style	"
	Old Farm Tools	Shire
	Vintage Farm Machines	"
	Wells & Water Supply	"
	Discovering Carts & Wagons	"
Wailes, Rex	The English Windmill	R.K.P.
Walton, James	Watermills, Windmills & Horsemills in South Africa	Struik

ACKNOWLEDGEMENTS:

The author would like to record his appreciation for the help he has received from the following individuals & institutions: Abbeydale Industrial Hamlet: American Museum, Bath: Gerald Anderson Collection: Armley Mills: Barbados Museum & Historical Soc.: Barbados National Trust: Lindsay Barnard: L.R. Blackmore: Bodleian Library: Borough of Beverley: Bouma Windenergie B.V.: Bourton-on-the-Water Motor Museum: Bucks. County Museum: Centre for Alternative Technology, Machynlleth: Central Electricity Generating Board: Chiltern Society: Alissandra Cummins: Country Life: Dept. of Environment: De Hollandsche Molen: Derrick Duddon: Electrum, Arnhem: Eling Tide Mill: Finch Foundry Trust: Stanley Freese: Forge Mill, Redditch: Roy Gregory: Peter Gunner, Sussex: Peter Lloyd Harvey: R. Harrison: John Haselfoot: The Harwich Society: Heatherslaw Mill: Peter Hill: R.S. Holmes: Hunday Museum: Inland Waterways Assn.: Anders Jesperson: Kaal-van der Linden: Kennet & Avon Canal Trust: Kirbee Collection: E.C. Kneale: Margaret & Fred Laverack: Llywernog Silver-Lead Mine: Lynton Cliff Railway: Donald Macloed, Shawbost School: Michelham Priory: Barbara Mortimer: Museum of East Anglian Life: Muncaster Mill: National Trust: National Trust, Northern Ireland: Museum of Antiquities, Scotland: Netherlands Energy Research Foundation, Petten: Netherlands Open Air Museum, Arnhem: Netherlands Tourist Office: David Nicholls: Norfolk Windmills Trust: North of England Open Air Museum, Beamish: Chris. & Robyn Nunn: Oxford House Industrial Archaeology Project: Oxford City Museum: Janet Peatman: Pitstone Local History Society: Polenko B.V., Rhenen: Raphael Salaman: Rosemary A. Scott-Smith: Scottish Development Dept., Ancient Monuments: Sealink: Caroline Simcoe-Gerson: Anton Sipman: Donald Smith: Sussex Archaeological Society: Sussex Industrial Archaeology Society: Fred Taylor: Textplan Ltd.: Nicholas Vince: Warwickshire C.C.~Architect's Dept.: Waterways Museum, Stoke Bruerne: Weald & Downland Open Air Museum: Woodbridge Tide Mill: Worsborough Mill: Irene Wright.

The author owes a particular debt of gratitude to his research assistant Barbara Vince; to Jacobus M. Stikvoort and to Roger Hudson.

Lakeland Cheese press from a painting by Robert Hills (1769-1844). Birmingham Art Gall. Note the lever to adjust the counterweight.

See 'OLD FARMS' p.26.

INDEX :

FORGE MILL · REDDITCH · p. 86.